Praise for
GOD SPARK

A Journey into the Light You Were Born to Embody

God Spark is more than a book—it's an awakening. It gently but powerfully invites you to remember who you truly are and to reclaim the light and truth already pulsing within your cells. It blends spiritual wisdom with scientific clarity in a way that feels both cosmic and deeply personal. Every page felt like a transmission of love, alignment, and purpose. If you've ever sensed you were meant for something more—but couldn't quite name it—*God Spark* will ignite that knowing within you and show you how to live from it, beautifully and boldly.

—**Kim Kiyosaki** CEO & Co-Founder RichDadPoorDad, Rich Women, Philanthropist, Investor

I speak as one who walks with the spirits of the canyon, who listens to the wind that carries the voices of the ancient ones. *God Spark* is not just a book—it is a remembering. Dr. Trish has touched something sacred, something we have guarded for generations: the knowing that the Divine lives within all things. Her words carry the rhythm of the

Earth and the breath of the Creator. She brings together the science of the modern world with the sacred codes of Spirit, and in doing so, she honors the original teachings that were sung before time began. This book is a gift—not only for today's people, but for the future generations who will need to remember the light within.

—**Uqualla,** Medicine Man, Shaman, Havasupi Tribe, Grand Canyon, Spiritual Emissary

I've known Dr. Trish for many years and have witnessed her unquenchable thirst to understand the life and teachings of Jesus—not just from a place of

belief, but from deep, lived devotion. *God Spark* is the radiant fruit of that journey. To me, this book captures the essence of what Jesus came to teach: that the Kingdom of God is within us, and that Divine alignment is not earned—it's remembered. Through her unique blend of scientific insight and Spirit-filled clarity, Dr. Trish offers a path back to truth, love, and the light we were born to carry.

—**Dr. Deborah Schlosser,** D.C., R.N, Professional Encourager, Human Relations Consultant

This book feels like the merging point between star maps and soul codes. In *God Spark*, Dr. Trish effortlessly translates the language of Quantum Science, Sacred Geometry, and Divine Intelligence into a compass for living. She shows why so many of our previous paths fell short—not because we failed, but because they weren't aligned with our true energetic design. With stunning clarity and brevity, she guides us into coherence with the frequency of Divine Order—where miracles are no longer the exception, but the natural outcome of alignment.

—**Laura Clancy,** Founder Muffin Toppled, World Champion Powerlifter, Broadway Singer, Actor

What I love most about *God Spark* is how Dr. Trish does what only a true master can do—take something vast, mysterious, and deeply spiritual, and make it simple, clear, and immediately usable. This book is short, but it carries the weight of lifetimes of wisdom. She expertly distills what others stretch into hundreds of pages into a concise, powerful message that hits home fast. It's not just a book—it's a shortcut to alignment, clarity, and activation.

—**Dr. Kelly Hannigan,** RN, NMD, Medical Aesthetics Expert

God Spark is a holy revelation wrapped in modern clarity. Dr. Trish speaks with the authority of one who walks closely with the Spirit, and yet she makes the mysteries of God's design feel intimate and accessible. With each page, she gently reveals why the world's way often fails—and how the Spirit's way leads us home. This book is not just a read; it's a return. A return to Divine order, sacred alignment, and the knowing that God is present in every breath, every cell, every step we take.

—**Cristina Davison**, CEO & Co-Founder, Razzle Dazzle, Inc. Redefining Luxury

Dr. Trish has an unmatched gift for cutting through noise and delivering truth in a way that lands with both precision and grace. *God Spark* is the essence, the distilled gold, the "just tell me what works" manual for waking up to your divine design. She simplifies the mystical without stripping its power, and what you're left with is a book you'll read and feel in your soul for years.

—**Jack Canfield**, Success Expert, Guinness World Record Best-Selling Author, Chicken Soup for the Soul series, Inspirational Speaker

God Spark speaks directly to the seeker, the leader, the healer within by bringing forth and expanding upon ancient wisdom and scientific knowledge. To quote the shamans, "Knowledge is understanding that water is H20 and wisdom is being able to make it rain." The compilation offers practical insights to live from the inside out and a path to realignment, peace, and power, without fluff or overwhelm. Be inspired to embrace joy and true purpose by accessing the energy of creation itself.

—**Melani Walton,** Global Philanthropist & Human Performance Specialist.
International Leadership Awardee
and Woman of the World recipient.
mutli-sport All-American athlete & scholar

GOD SPARK

"A Journey into the Light You Were Born to Embody"

DR. TRISH SANDERS

STARSHINE

EDUCATION RESOURCES, LLC
Phoenix Arizona

Published by: Education Resources, LLC Phoenix, AZ 85015

Cover Design: King of Designer

Disclaimer – Sacred and Real:

The author of this book does not dispense medical advice or prescribe the use of any technique as a form of treatment for physical emotionsl, or medcal problems without the advice of a physician either directly or indirectly. The intent of the author is only to offer information of a general nature to help you in your quest for emotional, physical, and spiritual well-being. In the event you use any of the information in this book for yourself, the author and the publisher assume no responsibility for your actions.

This book isn't for everyone. Some will love it. Others won't. That's okay.
It's not medical advice. It's not religious doctrine.
It's a remembrance — a frequency transmission — a spark.
Take what resonates. Leave the rest.
Trust your stirring.
Miracles happen when you're ready, not when you're forced.
You don't need fixing. You need remembering.
The angels are on my side. But they're on yours too.
Let the light guide you, not fear.
You are sacred. You are sovereign.
You are the spark.

The Library of Congress has cataloged this book

Names: Sanders, Dr. Trish, author

Title: **God Spark**: **"A Journey into the Light You Were Born to Embody"**

Subjects: Religion & Spirituality ' Religious Studies ' Psychology Self-Help ' Personal Transformation, Spiritual

Identifiers: ISBN: Tradepaper ISBN 978-0-9826133-7-5 Digital 978-0-9826133-6-8

Also By

DR. TRISH SANDERS
Aka TRISH MCCARTY

Strategic Partners in the E-Commerce

Totally Rich

Just Plane Yoga

StarShine Field School Teacher Training Manual

The StarShine Effect

Think and Grow Rich Today

Transform Your Life, Business and Health

Soul of Success with Jack Canfield

Dedication

To my children and grandchildren—
your lives are my greatest prayer
and proof that love ripples through generations.
You are the living sparks of my soul's light,
the continuation of this sacred remembrance.

To the Divine Spark within every living being—
the breath, the light, the truth, and the eternal remembering.
To those who have walked before us, holding the flame.
To those awakening now, with trembling hands and wide eyes.
And to those yet to come, who will rise in radiant remembrance.

This book is for you.
This spark is ours.

Preface:
The God Spark Is Real

This book began with a dream and was completed through divine orchestration.

It started years ago with a vision I received while living in a remote desert. The dream was simple and stunning: that every person, like every structure, carries Zero Point Energy inside of every atom—a limitless divine force waiting to be remembered, activated, and embodied. It is the stuff Einstein and Tesla spent their lives trying to figure out. That dream became a calling. That calling became this book.

This book was completed in a place just as sacred as its origin: the home of Bruce Allen, founder of the Taos Drum Company, and the land where my former husband, Steve McCarty of the Steve Miller Band, now lives. At the base of the Sacred Taos Mountain, a current of ancestral energy pulses—what locals call the *Taos Hum*. On the final night of writing, I was awakened repeatedly at 11:55, 2:55, 3:55, and 5:55 a.m. by what I first thought was a humming sound and live drumming and indigenous singing outside my window. But as I rose to listen, I realized the sound was coming from within—a vibrational remembering echoing deep inside my body and soul.

This is the God Spark: not an idea, but a frequency. A return to your original design.

This book is not meant to be read and forgotten. It is a mirror, a map, and a medicine. It is my offering to the Great Awakening unfolding on this Earth.

Welcome back.

XOXO DR. TRISH

Contents

Preface: *The God Spark Is Real* *11*

SECTION ONE: Remembering the Spark, Clearing the Static and Rediscovering Divine Truth 20

Introduction **21**

Chapter 1: Clearing the Static: Letting Go of False Beliefs 25

The Veil of Misperception 25

The Illusion of Separation 26

We Have Been Duped 28

Ho'oponopono Prayer for Clearing 29

The Emotional Inflammation Link 30

Forgiveness As a Frequency Reset 30

Environments That Connect, Not Divide 32

Chapter 2: The Science of the Spark 35

A Note on the Science of Remembering 35

Zero Point Energy and Quantum Coherence 37

The Schumann Resonance – Earth's Heartbeat 37

The Role of Belief in Bioenergetic Rewiring 39
Repatterning the Limbic Brain and Reclaiming Inner Safety 40
Neuroplasticity: You Can Change Your Mindset—and Your Mind 40
Maslow's Hierarchy: A Map for Healing and Ascension, Climbing the Sacred Pyramid: From Survival to Self-Realization 41
The Gap vs. Growth Mindset 42
The Sacred Climb 43
Reprogramming Your Inner Operating System 44
From Safety to Sovereignty 44
God Present in Every Cell 45
The Body as a Bioenergetic Transmitter 46
Negative Ions: Nature's Restorative Charge 46
The Connection Between Frequency, Synchronization, and Manifestation 48
Synchronization Practice 49
Scripture • Science: The Law of Synchronization 50

Chapter 3: Healing Humanity's Deep Wound 51

Separation as Humanity's Deepest Wound 51
Sacred Geometry and Unity Fields 52
Imagination as a Bridge Back to Oneness 54
Walking the "Yellow Brick Road" Within 55
How Collective Consciousness Shifts When Individuals Remember Truth 56
Chapter Reflection 57

SECTION 2: Aligning with Truth: Bringing Body, Mind, Spirit Into Frequency Harmony 58

Chapter 4: The Body as a Temple: Healing Emotionally and Physically 61
Emotional Release and Physical Healing Connection 62
Inflammation as Unprocessed Emotion 63
Honoring the Body as the House of the Soul 64
The Role of Beauty, Sacred Space, and Feng Shui in Coherence 65
Chapter 5: Food, Water, Fasting, and Detoxing 67
Fasting for Clarity and Cellular Renewal 69
Preparing for a Fast 70
Breaking the Fast 71
Structured Water and the Power of Intention 72
High-Frequency Food Compass for Vitality 74
The Frequency Spectrum of Foods 74
The Frequency of Illness and Health 75
Chapter 6: Sacred Movement, Color, and Joy: Activating Vibrancy 79
Movement as Medicine 80
Sacred Numbers, Cycles, and Rhythms 81
Color as Frequency Therapy 85
Clothing the Body in Frequency: The Hidden Power of Fabrics 87
Tips for Daily Life 89
How Fabric Frequencies Are Measured 90
Key Findings 90
Dance, Joy Practices, and the Dopamine Effect 91
Using Music to Boost Frequency 92
Sound and Color Synergy: The Frequency of Joy 94
Daily Joy Frequency Ritual 96

Chapter 7: Sacred Sexuality and Life Force: Honoring the Divine Energy of Creation 99

Kundalini and Sexual Energy as Divine Creative Force 101

The Sacred Body Knows. It Remembers. And it Responds. 102

Every breath becomes an act of holy creation. 103

The Biological Mirror 104

Sexual Wholeness 106

Chapter 8: Rituals, Breath, and Frequency Awakening the God Spark in Your Cells 109

God Spark Morning Frequency Flow 110

Bring the Healing Home 111

Breath Practices for Energy Activation 112

Sound as a Tool to Stimulate the Body's Natural Healing Systems 114

Your Voice as Medicine: Humming, Chanting, Mantras 115

Ancient Wisdom & Modern Frequency 115

God Spark Sound Practice (3–6 minutes) 116

Sacred Practice: God Spark Sound Reset (3–6 minutes) 117

SECTION 3: Embodying the Lighthouse: Becoming a Beacon 118

Chapter 9: Emotional Mastery and the Untethered Soul: Releasing Attachments 121

How Attachment Drains Frequency 122

The Biology of Attachment 123

The Spiritual Cost 123

Practical Tools for Emotional Detachment Without Disconnecting from Love 124

Chapter 10: The Magnetic Effect of High Frequency 127

Resonance vs. Effort in Manifestation 128

Why Alignment Attracts Faster Than Striving 129
The Physics of Resonance and Divine Order 130
The Body as an Electromagnetic System 130
Living as a Lighthouse Instead of Chasing Ships 130
The Power of Becoming Rather Than Pursuing 131

SECTION 4: Igniting the World: Spreading Light, Love, and Truth Through Action 132

Chapter 11: The God Spark in Relationships and Sacred Love 135
Holy Love, Divine Union 135
The Chemistry of Sacred Love 136
The Frequency of Sacred Love 137
The Fusion Point: Chemistry Meets Frequency 138
Cultivating Sacred Love as Spiritual Practice 138
The Difference Between Attachment and Divine Union 139
Understanding Attachment: The Illusion of Possession 139
Embracing Divine Union: The Sacred Mirror 141
Shedding Attachment, Stepping into Union 143
From Earthly Bond to Cosmic Dance 144
A Final Reflection 145
Soul Partnerships as Accelerators of Divine Purpose 146
Closing Blessing 149

Chapter 12: Purpose, Prosperity, and High-Frequency Wealth 151
The Invisible Laws of True Wealth 152
Tesla, Ether, and the Hidden Field of Supply 153
The Living Metaphors of Nature 153
The Role of Gratitude, Joy, and Coherence 154

Adamantine Particles and the Cosmic Currency of Light 155

Why "Push" Energy Blocks Abundance 156

The Role of Joy, Service, and Gratitude in Wealth Creation: Sacred Prosperity Principles 157

Chapter 13: Restoring Harmony with Earth and Cosmos: Forgotten Currents, Ley Lines, Love Grid Lines, Chakras 161

The Sacred Conclusion – Why This Chapter Ends the Book............161

Healing the Earth's Energy Grid: Ley Lines and Sacred Geometry 162

Love Grid Lines and Planetary Ascension Currents............................ 163

Human Chakras and Earth Chakras: A Mirror of Divine Design.........164

Longitude, Latitude, and Star Alignment: Earth's Position in the Cosmic Symphony 165

Ancient Tools, Living Intelligence – Divining Rods and the Invisible Currents of Earth... 165

Migration, Memory, and the Cosmic Compass 166

Practical Restoration: Harmonizing the Grid Within and Without.....168

A Call to Sacred Stewardship... 170

Final Activation: Sacred Sites, Healing HeARTS Love Grid Prayer.....171

STARSHINE..172

About the Author 175

SECTION ONE:
Remembering the Spark, Clearing the Static and Rediscovering Divine Truth

GOD SPARK

"A Journey into the Light You Were Born to Embody"

Introduction

Awakening to the Spark

Written, now, in sacred alignment at 1:11 PM on Easter Sunday Eve. Candles lit. Jesus candle glowing. Abundance and Joy oil anointed. Sacred Mantras playing in the background. Magic wand in hand, I feel spirit all around me.
This is where the true birthing began.

Awakening

There are moments when the veil thins—
when time stands still,
and we remember who we truly are.

The Moment Everything Shifted

It happened in a moment I wasn't expecting—
not during prayer, or meditation, or ceremony—
but in the stillness of the night, when everything was quiet.

This book came to me during one of those moments.
At 3:33 a.m. on the Summer Solstice, my fourth day of a three-day water-only fast. I was awakened by a force I can only describe as Divine Light. Not external, but internal—vibrating through every cell of my being. It was a presence so holy, so clear, that I knew I had to obey.
The words that poured into my being were not mine —they were a transmission, a sacred scroll from the higher realms. I wrote for hours, laughing, crying, fully immersed in what felt like Divine instruction.

This book is not here to teach you something new.
It's here to help you *remember.*

The message was this:

The Spark of God lives inside every human being.

It cannot be destroyed.
It is whole.
It is holy.
And it is time to remember.

Remember that you were never separate.
That the Divine has always lived inside you.
That every single cell of your body is encoded with the God Spark—a frequency of perfect truth, joy, healing, power, and peace.

You may have forgotten…
But the Spark never left.

This book is a homecoming.

It will clear the static.
It will realign your frequency.
It will invite you to rise.

Because you are needed—not someday, but now.
This is your moment to awaken.
This is your moment to shine.
This is your moment to remember.

I didn't hear words, but I *knew* them. They were alive in my body, dancing in my breath, pulsing with clarity.

I laughed. I cried. I was overcome with the sense that something ancient and sacred had just been returned to me. Not just for me—but for *us*.

You are reading this because your soul has been whispering:
"There's more."

There is.
And it begins now.

"God is within her, she will not fall; God will help her at break of day."

— **Psalm 46:5 (NIV)**

The Invitation

This book is not meant to entertain you.
It's meant to *awaken* you.
It's not here to give you more to think about—
It's here to help you remember what you already know.

You didn't stumble upon these words.
You were called here.

Maybe softly.
Maybe through pain.
Maybe through longing, or synchronicity, or a divine nudge that

whispered,
"There's something more."

This is your invitation to return.
To come back to the place inside you where truth never left.
To stop striving, fixing, or proving—
and simply **become**.

Chapter I:
Clearing the Static: Letting Go of False Beliefs

The Veil of Misperception

Before we can remember the Spark, we must acknowledge what caused us to forget. The world has taught you to question your worth, to doubt your light, to search outside yourself for answers that only God placed within you.

These are not accidents. They are intentional misperceptions, programmed into culture, religion, education, and even family systems— *not always out of malice, but out of misalignment.* Passed down like spiritual amnesia.

There are **five primary illusions** that cloud your inner divinity:

1. **The Illusion of Separation** – That God is somewhere else, and you are alone.
2. **The Illusion of Unworthiness** – That you must earn your value.

3. **The Illusion of Time** – That you are running out of it, or behind. Or simply in the wrong one.

4. **The Illusion of Control** – That safety comes from grasping instead of surrendering.

5. **The Illusion of Lack** – That there is not enough—of love, of resources, of you.

These illusions form a fog. A static frequency. They keep you chasing, comparing, doubting, and settling. They pull you into roles and identities that are *not your truth.*

But deep within—beneath it all—your Spark still glows. And the invitation is this:

"Come home."

The Illusion of Separation

What is one of the world's earliest domination mechanisms? "*Divide and Conquer"* a phrase that comes from the Latin divide et impera ("divide and rule") and it's an ancient political and military strategy. It has also become one of the most used marketing strategies to sell almost anything. It is even where the idea for "gaslighting" emerged. It is a strategy hard to detect when it is being used on you because it deeply effects our emotional well-being and understanding…on purpose.

Roman leaders, like Julius Caesar and later emperors used divide et impera, to control vast territories by splitting groups into small ones by geography, ethnicity, religion or interests. This prevented resistance. It was clever. In "The Art of War" Machiavelli kept potential rivals suspicious or in competition with each other so they couldn't unite against him. Brittain, Spain and other dynasties applied it worldwide to encourage division between tribes, classes and religious groups to maintain their control. In modern politics and society operating today in media, politics and corporate structures, they foster division so that people focus on fighting amongst one another instead of uniting to challenge the system in power.

From a spiritual and frequency perspective, this is exactly the "illusion of separation" I so often discuss. The "divide" keeps people in lower vibration (fear, distrust, competition), while "conquer" keeps power centralized in those creating the division. Humanity was purposely pulled away from unity consciousness. And even "unity" consciousness was misconstrued to produce communism and socialism, just another radical example of ultimate control by a few.

But this "illusion of Separation" is more embedded than we have thought to consider. Think about our human body and our choice of healing modalities. Physicians and healers were once sought out to look at our whole body, our environment and even our relationships within our tribe or families. They were taught the interrelationship between everything. Today physicians, nurses, hospitals and nursing homes are taught to specialize. Typically, healthcare workers do not look at the relationship between one part and another. Eastern Traditional Chinese Medicine, acupuncture, and 5000 year old, Ayurvedic Medicine, assess an interrelationship between emotion, environment, spiritual and various body parts but they too are limited in their practices.

For seventeen years, I worked to revise curriculum for K-12 children into a more holistic way to learn and experience learning. I was shocked at how we teach our children to separate themselves, their bodies, their classmates and even their motivations.

Even architects, unknowingly, have been responsible for creating division among families, neighborhoods and communities. I was told once by a Frank Lloyd Wright Fellow, the beginning of the "white picket fence" was the beginning of the fall of America. Perhaps it was in a moment such as that, I became even more impassioned to teach the unity of our own body, mind, spirit, health, wealth and happiness. It is important now, for all of us to help reunite our communities to one another and all of us to mother earth, the cosmos and to the teachings of Jesus.

We Have Been Duped.

Somewhere along humanity's journey, a false story was planted and repeated so oten that we began to live as if it were truth. The story says we are separate—separate from each other, separate from the Earth and separate from God.

We have been convinced that life is a competition that resources are scarce, that we must protect ourselves from one another. We have been conditioned to see "the other" instead of "another me." We have been conditioned to blame, shame, judge and condemn others rather than looking at our own responsibility in creating the separation or fear.

At the level of frequency, there is no gap between us. Science confirms what the mystics have always known, all matter, all life, all consciousness is woven together in one unified field.

When we awaken to this, the game changes. We no longer play by the rules of division. We remember that every act of love, every moment of gratitude, every choice to forgive is a direct strike against the lie of separation. This is the law of physics and considers the prayer of idealism.

We have been duped…but we do not have to stay that way.

Ho'oponopono Prayer for Clearing.

When the veil of separation begins to dissolve, old thoughts, fears and judgments start to rise to the surface. These are the static signals that distort the truth of who we are. One of the simplest and most beautiful tools I have found for clearing this static is the Hawaiian practice of Ho'oponopono.

It is a prayer in its pure form, Ho'oponopono is a way to restore harmony – with ourselves, with others, and with life itself. It is not a long process or a complicated ritual. It works at a cellular, biological level. It causes a frequency shift. It causes your brain to rewire.

I use it as soon as I notice my heart closing or my mind pulling into judgement. I simply speak the four phrases, directing them toward myself, another person, or even the situation itself. I usually don't tell the other person I am doing it. I repeat the phrase over and over in my mind until I feel my body shift. I am basically telling my body I am sorry for not seeing things differently:

> ***I'm sorry.***
> ***Please forgive me.***
> ***Thank you.***
> ***I love you.***

Within seconds, my energy changes. The separation softens. The veil thins. I am back in connection. This is not the same as the deep, life-changing work of forgiveness we will explore later – but it is a gentle key that can unlock the door to that deeper healing.

The Emotional Inflammation Link.

When we believe we are separate – from others, from nature, from God – the body responds as if it's in constant danger. This state of disconnection triggers a quiet storm inside us: stress hormones surge, muscles tighten, the heart beats faster, and our cells prepare for defense.

Science now confirms what ancient wisdom has always known: chronic fear, judgement, and isolation do not weigh on the heart – they inflame the body. The very chemistry of separation fuels inflammation at a cellular level, clouding our thinking, draining our energy, and dimming our light.

The good news? The moment we choose connection – a kind word, a moment of gratitude, a breath of peace – the body's chemistry begins to shift. Inflammation cools, Energy returns. The God Spark shines a little brighter.

Forgiveness As a Frequency Reset.

Forgiveness is more than an act of kindness – it is a radical frequency shift. When we hold onto resentment, betrayal, or regret, those emotions create a heavy, low-vibrational field around us. We may not even realize it, but our very cells are resonating with the memory of the wound.

When we choose forgiveness – not to condone the action, but to free ourselves from the energetic chain – something extraordinary happens. The emotional weight dissolves. The static clears. Our frequency rises.

Forgiveness recalibrates the entire energy field. It's like clearing the interference from a radio signal: suddenly the broadcast is pure, strong, and aligned with truth.

The Ho'Oponopono Prayer can help in the moment. But a deep act of forgiveness, perhaps an understanding and acceptance is a complete reset – a liberation that changes the frequency you live in every single day.

Holding onto resentment is like carrying a heavy stone in your pocket everywhere you go. At first, you barely notice it. But over time, it changes the way you walk, the way you stand, the way you breathe. It's always there, pulling you your energy, altering your rhythm.

Forgiveness is the moment you reach into that pocket and set the stone down. The instant you let go, your whole-body shifts. Your shoulders lift. Your stride changes. You can breathe deeply again.

Nothing outside you has changed – the past event still happened – but you have changed. The weight is gone. Your frequency has risen to match the lightness you now feel. Some people say holding onto bitterness is like drinking poison hoping it will hurt the person you are resenting.

It does take effort to forgive. We have not been taught tools with which to know how to go about forgiving. We will get deeper into this subject in other parts of this book. But for the shorter version, it is not the act but your judgement, belief or story around the act that has caused you to suffer. If you can take the responsibility to change the story you believe, you will begin your healing journey. If you can tell yourself that if you were that person, with the same upbringing, the same life circumstances, you perhaps would have done the same thing. In this way, you are not excusing their behavior, you are making sense of it in your own head and heart and you are seeing them as having disadvantages you have not had to endure. This is not ever easy, but it is always possible. Smelling certain Essential Oils will go through the blood-brain barrier to disrupt the emotional molecules impregnated with the chemicals of the memory of event and will help to release the feelings surrounding it.

Environments That Connect, Not Divide.

The environments we live, work, sleep , and choose to be in are not neutral. They hold memory, energy and frequency. A cluttered, stagnate, dirty space around us can reinforce the illusion of separation – from beauty, from nature, from abundance, from our Holy self, our own clarity and from others. Our spaces around us either draw us into connection or quietly reinforce our disconnection. They keep us feeling unfulfilled, unfinished, broken and resentful without us even realizing it. When they are sterile, colorless, disconnected from nature, these spaces remind us that everything is disconnected, separate, mechanical, not alive.

Just as we clear emotional static, we must also clear environmental static. This is not a reminder to be more tidy. This is a reminder about how space shapes our spirit and impacts our day to day existence of happiness, abundance, expectation, wellness, love and faith. A heavy, chaotic environment keeps us in survival mode reinforcing the illusion we aren't safe, we aren't worthy of beauty and orderliness, of fitting with others or ourselves. Most of us live in environments shaped by a culture that values speed, consumption, excitement and dopamine rather than freedom, harmony, belonging and service to self and others.

Clearing out our environment is one of the fastest ways to dissolve the illusion of separation, build confidence and feel harmonious. This could be as little as opening a window, letting in sunlight and fresh air. It could also be very meaningful; establishing an alter space for your prayers, reflections, remembering, and creating with a flower, a candle, a picture of Jesus, a picture of love or anything that helps you to connect with your soul and Higher Self. This Sanctuary area has been a very important part of every home, every office and even every hotel room I've spent the night with.

In Chapter 4, we will take a deeper exploration of Feng Shui, clearing clutter and the placement of order. I am reminding you of the simple daily practice, and developing the habit, of bringing your spaces into the coherence that supports you, so it reflects your truth and not the lie you have believed. When your environment holds the frequency matching peace and connection, your own frequency will automatically rise to meet it.

Chapter 2:
The Science of the Spark

A Note on the Science of Remembering

Scientists have shown that the human brain is plastic—it reshapes itself constantly based on thought, belief, and emotional frequency. When you begin to say **yes** to peace, truth, love, and acceptance; your brain begins to build new neural pathways to match that frequency.

It's spiritual. And it's biological.

Each act of sacred remembrance is also a biological reprogramming.
Your nervous system is learning that it is *safe* to shine.
Safe to love.
Safe to remember what you've always known.

Your **yes** is not just a prayer.
It's a command to your cells:
"Return to light."

James Murray Walker, a healer I have the blessings to know and work with, once told me the story of **K'garnea**. It was passed down by the monks of the high Himalayan mountains, where the air itself carries the frequency of silence and surrender. Their teacher, **Saipal**, often reminded them that true healing never comes from force, effort, or even knowledge—it comes when we unclench the grip of control and let the Divine flow through us.

"K'garnea," translated into English, is much like the phrase *"Let go and let God."* It became a sacred saying among the people of Yangar, in the Humla province near the northwestern border of Tibet in Nepal—one of the harshest, most difficult environments on earth, where daily survival demands both resilience and faith.

Saipal embodied this teaching so completely that he became a living parable. His presence carried no resistance, only trust. To those around him, he often repeated the simple truth: *"K'garnea—let go, and let God."*

James explained that this teaching was not passive resignation but the most profound act of courage. To let go means aligning oneself with the unseen order of the universe, trusting that every cell, every breath, every unfolding moment is supported by a greater intelligence.

Even Albert Einstein, who spent his life mapping the fabric of reality, arrived at the same realization. He once said the most important question we can ask is: *"Is the universe friendly?"* His answer was a resounding yes. The universe, he believed, is always for us—never against us.

And here is the mystery revealed: whether you believe the universe is for you or against you, you will prove the truth.

Placed side by side, the wisdom of the monks and the clarity of science offer the same invitation: surrender into trust. Healing, wholeness, and even genius awaken not through grasping but through allowing.

This is what James has lived and practiced as a healer, and now has become one of my own foundational sayings/reminders. The story of K'garnea is not ancient myth—it is alive in those who, like James, embody the courage to step aside so that the Divine can move freely through them.

Zero Point Energy and Quantum Coherence

Quantum physicists now describe what they call the **Zero Point Field**—a state of absolute stillness beneath all form and motion. Even in a perfect vacuum, energy never disappears; it remains as pure potential. The Zero Point is, in truth, a vast ocean of infinite, intelligent energy that underlies everything we see and experience.

Long before the term *Zero Point* existed, **Nikola Tesla** intuited its presence. He believed this invisible energy could be harnessed freely and demonstrated it in Colorado Springs, lighting bulbs without wires by resonating with the Earth's natural frequencies. **Albert Einstein**, through his theory of relativity, described a unified field connecting all matter—an elegant, subtle order behind the visible world. Both visionaries were reaching for what you, in this moment, are remembering: **the God Spark**—the vibrational core of all life.

The Schumann Resonance – Earth's Heartbeat

In **1952**, German physicist **Winfried Otto Schumann** mathematically proved that the space between the Earth's surface and the ionosphere acts as a resonant cavity, vibrating like a giant drum. He predicted that this cavity would oscillate at a fundamental frequency of about **7.83 Hz**. Later, in **1960**, researchers were able to measure it directly and confirm its existence.

This frequency, now called the **Schumann resonance**, is sometimes referred to as the *heartbeat of the Earth*. It falls within the same range as **human brain waves** (theta and alpha states), the states linked to meditation, creativity, and healing.

Its importance goes beyond theory:

- **NASA and space agencies** discovered that astronauts cut off from this resonance in orbit developed insomnia, anxiety, and lowered immunity. To protect them, spacecraft were equipped with **artificial Schumann resonance generators** so their bodies could stay in sync with Earth's natural rhythm.
- **Farmers and agricultural researchers** in the 1970s (notably through **Dan Carlson's "Sonic Bloom" method**) began playing recordings of **birdsong** alongside foliar feeding, finding that plants opened their stomata wider, absorbed more nutrients, and grew stronger and more abundant.
- Mystics and sound healers also note that the **frequency resembles the hum of bees**, nature's harmonizers, whose collective vibration supports pollination and plant growth.

From snakes sensing earthquakes, to bees humming, to birds guiding crops, all of nature confirms this truth: **life is tuned to frequency.** The Schumann resonance reminds us that human beings are designed to live in coherence with the planet itself. When we align with Earth's natural pulse, we awaken the *God Spark* in every cell.

The Role of Belief in Bioenergetic Rewiring

Belief is not merely a psychological state; it is a bioenergetic command. When we believe something deeply, that conviction cascades through the nervous system, activating chemical messengers, reprogramming neural circuits, and influencing the body's electromagnetic field. In this sense, belief acts as both blueprint and builder—guiding the reconstruction of inner pathways that shape how energy flows through our cells.

Modern neuroscience confirms that expectation alters perception: the brain does not passively record reality, it filters and edits reality to match belief. In quantum biology, this extends further—belief creates the conditions under which certain energetic patterns stabilize or dissolve, determining whether coherence or chaos dominates within the cellular environment.

From a spiritual perspective, belief can be understood as alignment with resonance. To hold faith, to trust, is to tune the instrument of the body into frequencies that sustain vitality. Negative or limiting beliefs, conversely, create static—interrupting the body's ability to receive and broadcast its natural coherence.

Thus, bioenergetic rewiring is not only a process of changing thought—it is the embodiment of a new belief at the cellular level. Every healing, every transformation, begins with the courage to adopt a new inner agreement about what is possible. Belief is the invisible switch that changes the current.

"Belief is the frequency that shapes reality; it builds every miracle, every downfall, and every resurrection."

Repatterning the Limbic Brain and Reclaiming Inner Safety

We now know, both through ancient wisdom and cutting-edge neuroscience, that **what we believe shapes what we perceive—and what we perceive becomes our reality**. This is not just philosophy. It's biology.

The **limbic brain**, often called the emotional brain, governs your instinctual responses: fear, safety, connection, and survival. It evolved to **scan for threats**, store trauma, and activate protective patterns. In today's world, this can look like chronic anxiety, emotional reactivity, avoidance, or even overachievement—all strategies wired for safety, not necessarily truth.

The limbic brain is not logical. It's not concerned with whether your belief is real. It only asks:

"Is this familiar? Is this safe?"

That's why **limiting beliefs**—even ones we no longer consciously agree with—can still live deep in the body. When repeated often enough, a belief becomes a **neural pathway**. Like a well-worn trail in a forest, your thoughts follow the familiar route—even if that route leads you in circles of fear, lack, or self-doubt.

But there is hope.

Neuroplasticity: You Can Change Your Mindset—and Your Mind

Thanks to **neuroplasticity**, your brain can form new neural connections throughout your life. You are never too old or too wounded to rewire yourself. With every sacred yes to truth, love, and alignment, you are reshaping your brain, chemistry, hormones, and even your DNA expression.

This is the foundation of **epigenetics**—the science that proves your environment, thoughts, and emotional frequency affect which genes turn on and off.

Maslow's Hierarchy: A Map for Healing and Ascension. Climbing the Sacred Pyramid: From Survival to Self-Realization

Understanding how to reprogram your body and mind begins with recognizing where your nervous system is operating most of the time. Dr. Abraham Maslow's Hierarchy of Needs provides a powerful framework for this. Often depicted as a pyramid, it begins with survival at the base and ascends toward self-actualization at the top. It's not just a psychological model — it mirrors energetic frequency states in the body.

The Pyramid Structure:

1. **Physiological Needs** (food, water, rest, safety)
2. **Safety and Security** (shelter, stability, protection)
3. **Love and Belonging** (relationships, connection, tribe)
4. **Self-Esteem** (confidence, identity, respect)
5. **Self-Actualization** (purpose, truth, inner spark)

Most people cycle through the **bottom three layers** for much of their lives, especially when their **limbic brain** is constantly scanning for threat. This ancient part of the brain evolved to protect us — but it's also wired to expect fear, danger, and even death. It responds more quickly to negative signals than positive ones, which means the body can stay trapped in survival mode, even when the external environment is safe.

The Gap vs. Growth Mindset

When stuck in survival, people tend to live in a **gap mindset** — focused on what's missing, what's wrong, or what could go wrong. It's a lens of not-enoughness. Gratitude, creativity, and higher vision become inaccessible. But when we shift into the **growth mindset**, we rise above the limbic brain and allow the prefrontal cortex — the visionary, planning, and soul-connected part of our brain — to come online.

When fear runs the show, we fall into what could be called **"Gap Mindset"**—a pattern of seeing what's missing. You may feel like:

- You are "behind"
- You're not enough
- There isn't enough time, money, love, or space
- Everyone else got the map but you

This mindset comes from **limbic looping**—a nervous system stuck in the frequency of lack.

In contrast, a **Growth Mindset** acknowledges that healing, learning, and joy are possible—even now. It says:

"I may not have arrived yet, but I'm on the path."
"My past shaped me, but it does not define me."
"Every belief can be upgraded when I remember who I am."

This shift can be supported through breathwork, essential oils, meditation, nature, grounding, prayer, and light. Over time, we begin to

reprogram the brain's fear filters and create new neural connections based on love, possibility, and peace. When your nervous system learns it is safe to grow, your soul begins to stretch.

The Sacred Climb

Each step up the pyramid reflects a shift in energetic frequency:

- From survival to trust
- From fear to safety
- From isolation to connection
- From shame to confidence
- From confusion to clarity
- From longing to fulfillment

You don't climb the pyramid by force. You ascend through **alignment**. When your internal and external world reflect safety, love, and truth — you rise. And at the top, you remember:

You were never broken. You were only climbing home.

You **cannot ascend the pyramid without addressing what lives beneath it**.

A traumatized nervous system is not concerned with purpose.
A hungry body cannot meditate itself into joy.
A heart still expecting abandonment cannot fully open to love.

That's why the **healing journey is sacred**. You are not broken—you are **returning to safety**. Only from safety can your nervous system relax, your limbic brain loosen its grip, and your soul begin to lead.

Reprogramming Your Inner Operating System

To shift out of survival and into sacred creation, we must create new **loops of truth**. These can include:

- **Daily affirmations spoken aloud** while touching the heart or belly
- **Sacred breathwork** to signal safety to the nervous system
- **Essential oils and touch rituals** to create new body-based associations
- **Journaling, visualization, or inner child repair** to upgrade outdated narratives

Every time you choose faith over fear, truth over trauma, your limbic brain learns:

"I am safe. I am loved. I can expand."

Your nervous system doesn't rewire through force. It rewires through **consistency, compassion, and coherence**.

From Safety to Sovereignty

When your inner world feels safe, your biology begins to reflect it. When your beliefs align with divine truth, your energy field becomes magnetic.

And from this sacred space, your **God Spark** shines—not because you fixed yourself, but because you finally felt safe enough to **be yourself**.

God Present in Every Cell

This Spark is alive in every cell of your being. It is encoded in your DNA, the divine blueprint of who you are. Modern research confirms that DNA is not static—it responds to sound, light, and emotion. In the emerging field of quantum biology, scientists have discovered that DNA both emits and receives photons—particles of light. You are, quite literally, made of light. And the frequency you live at shapes how that light is expressed.

The psalmist wrote, *"For You created my inmost being; You knit me together in my mother's womb. I praise You because I am fearfully and wonderfully made"* (Psalm 139:13–14). Scripture is not using metaphor here—it is describing an energetic and biological truth. The God who spoke the universe into existence with light also placed that same creative light within you.

Jesus affirmed this when He said, *"You are the light of the world... let your light shine before others"* (Matthew 5:14–16). He was not speaking only of kindness or moral goodness; He was pointing to the literal, divine radiance designed to emanate from your being. When you align your thoughts, emotions, and actions with the frequency of love, you allow this light to expand.

Science now catches up with scripture, confirming that your cells, when coherent and healthy, emit stronger, more organized light. This is the God Spark—alive in you from the beginning, waiting for your conscious agreement to shine without limit.

The Body as a Bioenergetic Transmitter

Your body is not just a vessel—it is a living, dynamic transmitter of energy and information. Every heartbeat, every breath, every movement sends ripples of frequency into the world around you. Just as a tuning fork can set another into vibration, your body resonates with the frequencies you embody—whether love, fear, peace, or chaos.

At the quantum level, your cells are in constant communication—not only with each other, but with the environment, the Earth, and even the cosmos. This communication happens through light (biophotons), sound (vibrations in the tissues and fluids), and electromagnetic fields generated by your heart and brain. In fact, the heart's electromagnetic field is measurable up to several feet away, and its rhythm can influence those nearby, often without a single word spoken.

When your inner frequency is coherent—aligned in body, mind, and spirit—you broadcast clarity and stability into the collective field. You become a beacon. This is why prayer, meditation, worship, and intentional movement are so powerful—they don't just shift your inner state; they alter the energetic landscape around you.

Your body was designed for this. It is both a receiver and a transmitter, a sacred instrument tuned by the Creator to carry the God Spark into the world. Every thought, word, and action sends a signal. The question is: what station are you broadcasting?

Negative Ions: Nature's Restorative Charge

Modern science has confirmed what ancient peoples instinctively knew: being near the elements of nature restores balance to the human body and mind. One of the key reasons is the presence of **negative**

ions—electrically charged particles that are abundant in natural environments where energy is moving or matter is transforming. Negative ions increase oxygen flow to the brain, stabilize mood, improve sleep, and strengthen immunity. They literally recalibrate the body back into coherence.

Ocean Waves

The breaking of waves at the shoreline releases a flood of negative ions into the air. This is why people often feel an immediate sense of calm, clarity, and uplift when walking on the beach or breathing in sea spray. The ocean is one of the most powerful natural ionizers on Earth.

Fire

Fire also generates negative ions as wood burns. For centuries, people have instinctively gathered around fires not only for warmth and light but also for the subtle medicine of its energy. The flickering flame and crackling wood soothe the nervous system, synchronize rhythms, and create a sense of peace and connection. Even a single candle can shift the energy in a space, which is why fire has been central in rituals, homes, and sacred ceremonies throughout history.

Forests and Essential Oils

The atmosphere of a forest is saturated with negative ions. In addition, trees and plants release aromatic compounds known as **phytoncides** (including terpenes such as alpha-pinene and limonene). Breathing these compounds, which are also present in therapeutic essential oils, especially the Young Living Oils I use for purity, lowers cortisol, boosts immunity, and clears the mind. This is why practices such as "forest bathing" have measurable effects on human health and well-being.

Autumn Leaves

Even the scent of autumn leaves carries this renewing energy. As leaves decompose, they release terpenes and organic compounds that create both fragrance and subtle electrical charge. The crisp air of fall, rich with this chemistry, evokes nostalgia, grounding, and a natural reset of the body's rhythms.

Ancient people may not have had the scientific language of "ions," but they understood the experience: the clarity of ocean air, the comfort of firelight, the vitality of the forest, the renewal of autumn. Modern science now confirms that each of these elements restores the body's natural coherence. When we immerse ourselves in them, we are not just connecting with nature—we are remembering the original harmony of creation itself.

The Connection Between Frequency, Synchronization, and Manifestation

Everything in creation moves to rhythm and resonance. From the spin of electrons to the orbits of planets, there is a natural synchronization woven into the fabric of the universe. Frequency is the language of this order—and synchronization is the moment when your personal signal comes into harmony with it.

Your frequency is the sum of your thoughts, emotions, beliefs, and physical state. It acts like a broadcast signal, attracting experiences and opportunities that match its tone. Synchronization happens when that inner broadcast aligns with the greater rhythms of divine design—what some call the flow, the unified field, or God's perfect timing. In this state, manifestation is no longer a struggle; creation moves toward you as effortlessly as the tide returning to shore.

Science offers glimpses of this mystery: fireflies flashing in unison across a field, heart cells in a lab beating together without any visible conductor, metronomes naturally falling into sync when placed on the same platform. Spiritually, this is what Jesus described when He said, *"Seek first the kingdom of God and His righteousness, and all these things shall be added to you"* (Matthew 6:33). The "kingdom" is coherence—perfect resonance with divine order.

Manifestation, then, is not about willing your desires into being by force; it is about tuning yourself to the frequency of the reality you seek, and letting synchronization pull it into form. When your God Spark is aligned—your heart steady, your mind clear, your spirit surrendered—you become an open channel for the unseen to become seen.

Synchronization Practice

1. **Pause** – Close your eyes, place your hand over your heart, and take three slow, deep breaths.
2. **Attune** – Whisper softly: *"I align my heart, mind, body, and spirit with the frequency of divine order."*
3. **Visualize** – See yourself standing in a stream of golden light, the pulse of the universe flowing through you, matching your heartbeat.
4. **Release** – Let go of tension in your shoulders, jaw, and hands, as if dropping all resistance into the earth.
5. **Receive** – Hold a single, peaceful thought of the reality you are calling in—without grasping, just allowing it to exist.

Do this daily for a few minutes, and you will begin to feel the subtle shift as your frequency synchronizes with the flow of creation.

Scripture • Science: The Law of Synchronization

Scripture

"Can two walk together, except they be agreed?"
— Amos 3:3

"For where two or three gather in my name, there am I with them."
— Matthew 18:20

These verses point to resonance—agreement of spirit, thought, and intention—as the place where divine presence and power flow most freely.

Science

In physics and biology, **synchronization** describes the phenomenon where oscillating systems naturally align their rhythms. Fireflies flash in unison, heart cells beat together without instruction, and pendulums swing in harmony. When systems synchronize, their combined energy becomes more efficient, stable, and amplified—just as in prayer, worship, or collective intention.

Takeaway

When your inner frequency resonates with the greater divine order, you are no longer moving alone—you are carried by the same current that holds galaxies in place.

Chapter 3: Healing Humanity's Deep Wound

From the moment the first human believed they were separate from the Source, a shadow fell across the collective heart of our species. This illusion—subtle at first—grew into a story told in every culture: the story of "me" and "you," "us" and "them." Over centuries, it hardened into systems, borders, and identities that divided rather than united. We forgot the original harmony that sings in every atom, the invisible threads that bind all creation in a single, living tapestry. Yet even as humanity fractured, the memory of oneness never left us. It lives in the way we long for connection, in the ache for justice, in the quiet recognition that beauty, love, and truth feel like home. Healing this wound is not about erasing our individuality—it is about remembering the deeper unity beneath it, the original design in which every soul is a facet of the same divine light

Separation as Humanity's Deepest Wound

Separation is more than a feeling—it is a distortion in perception that has altered the course of human history. Unlike personal disconnection, which can be mended through individual healing, this wound exists at a collective level. It is the inherited belief that we are isolated from the Divine, from one another, and from the Earth itself. This fracture has fueled conflict, justified exploitation, and left humanity in a constant state of longing for something unnamed.

Ancient wisdom traditions tell us that this wound began not with a single event, but with a gradual forgetting. As societies turned outward—toward survival, conquest, and control—the inward knowing of unity faded. Over time, we came to define ourselves through contrasts: race, class, nation, religion. These identities were meant to be threads in a diverse tapestry, but under the influence of the separation wound, they became walls.

The danger of separation is not only in what it divides, but in what it obscures. When we see ourselves as apart from life, we lose awareness of the living field that connects us—the same field that makes compassion instinctive, healing possible, and miracles natural. This disconnection starves the soul, leaving a hunger no amount of material achievement can fill.

Yet the wound is not beyond repair. In fact, the very pain it causes can become the signal that guides us back. Every moment of loneliness, injustice, or ecological crisis is an invitation to remember the truth: that beneath the illusion, we have always been one body, one breath, one heartbeat in the mind of God.

Sacred Geometry and Unity Fields

If separation is the wound, unity is the original blueprint—and it has never been lost. All of creation is built on repeating patterns of harmony, known across cultures as **sacred geometry**. From the spirals of galaxies to the unfurling of a fern, from the honeycomb's perfect hexagon to the chambers of your own heart, the same mathematical language is expressed again and again. This geometry is not decoration—it is the architecture of life itself.

These patterns reveal that nothing in creation exists in isolation. Every form is nested within another, each curve or line precisely proportioned to fit into the greater whole. Physicists speak of the **unity field**—the energetic matrix that links all matter and consciousness in a seamless web. Spiritual traditions describe it as the body of God, the net of Indra, the Tree of Life. Whatever the name, the truth is the same: there is no "outside" to the whole.

When you begin to see the patterns, you also begin to feel them. Standing beneath a star-filled sky or tracing the petals of a flower, something inside you recognizes the order. It is the same intelligence that spins your DNA into a double helix, that beats your heart in perfect time, that draws your breath without thought. You are not merely surrounded by these patterns—you are made of them.

Just as these patterns govern the architecture of the cosmos, they are also encoded in the human body. The proportions from chin to forehead, from navel to crown, from fingertip to fingertip mirror the same mathematical ratios found in seashell spirals, flower petals, and planetary orbits. These are not accidents of biology—they are expressions of the same harmonic laws that guide the distances between Earth and Mars, the timing of the moon's phases, and the pathways of the sun across the sky. Ancient builders, healers, and navigators understood this precision and used it to align temples, monuments, and living spaces with the greater unity field. By living within these proportions, whether in our own posture or in the spaces we inhabit, we step into resonance with the original blueprint of life.

Healing the wound of separation begins with remembering this truth: your very body is a piece of sacred geometry, vibrating in resonance with the unity field. And when you attune to this resonance, you step back into the flow that has always been holding you.

Imagination as a Bridge Back to Oneness

Imagination is one of humanity's most divine faculties. It is the ability to hold in mind something that does not yet exist in physical form—and by doing so, begin to bring it into being. Every invention, every movement for justice, every work of art began first as an image in someone's inner world.

But imagination is neutral. It can be used to create beauty or to reinforce separation. Left unconscious, imagination often turns into worry, fear, or disappointment—painting vivid inner pictures of what we dread. The body responds to these imagined scenes as though they are real, releasing the same stress chemicals that keep us locked in cycles of defense and disconnection. In this way, misused imagination becomes a tool for deepening the wound of separation.

Yet when imagination is guided by love, it becomes a bridge back to oneness. It allows us to see possibilities beyond the present limitations. The sculptor Michelangelo is said to have remarked, *"I saw the angel in the marble and carved until I set him free."* That is the essence of divine imagination—not forcing something into shape, but revealing the wholeness that was there all along.

Anthropologist Margaret Mead famously reminded us, *"Never doubt that a small group of thoughtful, committed citizens can change the world; indeed, it's the only thing that ever has."* Every such group begins with shared vision—people daring to imagine a healed world, and then taking aligned action to make it real.

Your imagination is a sacred instrument, capable of amplifying either the illusion of separation, creating intention to manifest your dreams and goals or the reality of unity. When you choose to imagine from the God Spark within, you are no longer daydreaming—you are transmitting blueprints into the unity field, where the pattern of wholeness can manifest.

Walking the "Yellow Brick Road" Within

The journey back to oneness is not a straight line—it's more like the winding yellow brick road. In *The Wizard of Oz*, Dorothy sets out on a long and colorful quest, believing she needs a great and powerful wizard to send her home. She meets friends along the way, faces storms, and braves uncertainty. Yet in the end, the Good Witch Glinda tells her the truth: *"You've always had the power, my dear. You just had to learn it for yourself."*

I once had my own yellow brick road moment, though mine began barefoot in an open field. I was walking without destination, feeling the warmth of the earth under my feet, each blade of grass bending and springing back as if in greeting. The wind moved gently through the trees, carrying a scent I couldn't name but somehow recognized. In that moment, there was no separation between my breath and the breath of the world. No striving, no reaching—just a clear, quiet knowing that I was home.

This is the deeper truth of the yellow brick road: it is not about arriving at some distant place or meeting a figure who will hand you the answers. It is about the steps you take within yourself, each one revealing the power, love, and wholeness that have been there all along. The companions you meet—like the Scarecrow, the Tin Man, and the Lion—are reflections of your own mind, heart, and courage awakening as you travel.

When you walk this inner path with bare feet, you remember what connection feels like in your body. You sense that every step you take is on sacred ground, every encounter a mirror, every breath an exchange with the Source of life itself. And like Dorothy, you realize the truth is not "out there." The God Spark has been inside you from the very beginning.

How Collective Consciousness Shifts When Individuals Remember Truth

When even one person awakens to the truth of their God Spark, the effect ripples far beyond their own life. Every coherent thought, every act of compassion, every moment of living in harmony with the unity field subtly shifts the vibrational landscape for everyone. Just as one tuning fork can bring others into resonance, a single heart aligned with love influences the frequencies of those nearby—often without a word spoken.

This is not poetic imagery; it is measurable reality. Studies in neuroscience and heart coherence show that when individuals enter states of peace, gratitude, or joy, their brain and heart rhythms become more ordered. This order radiates outward, affecting the electromagnetic field shared by all living beings. In the language of faith, your light is not hidden—it shines into the darkness, and the darkness cannot overcome it.

When enough individuals remember truth, a tipping point occurs. The stories of fear and separation lose their power. Systems built on exploitation begin to crumble. New patterns emerge—rooted not in control, but in collaboration; not in scarcity, but in the abundance that unity reveals. This is how revolutions of consciousness happen: quietly, steadily, as hearts wake up one by one until the whole field begins to glow.

You are part of that shift. Every time you choose alignment over fear, connection over division, truth over illusion, you help to heal humanity's deep wound. The yellow brick road you walk within is not yours alone—it becomes a golden thread in the great tapestry of awakening, guiding others home to themselves, and to the oneness that has been here all along.

Chapter Reflection

The wound of separation may be humanity's oldest pain, but it is also our greatest teacher. It calls us to remember that beneath every illusion of division lies an unbreakable unity—woven into the geometry of creation, sustained by the harmony of the unity field. Through imagination, we can choose to paint pictures of fear or to carve away all that hides the beauty already present. Like Dorothy on the yellow brick road, we discover that the answers we seek are not in a distant wizard's hand, but in the God Spark within us. And as each of us awakens to that truth, we add our light to the collective, shifting the consciousness of the whole toward healing, connection, and the joyful remembrance that we have never truly been apart.

Experience this short guided Ho'Oponopono Forgiveness Prayer and feel the shift immediately —

Just Scan Here

SECTION 2:
Aligning with Truth: Bringing Body, Mind, Spirit Into Frequency Harmony

GOD SPARK

*"A Journey into the Light
You Were Born to Embody"*

Remembering the God Spark is only the beginning. Once we awaken to unity, the next step is to bring that truth fully into our lived experience—into the way we think, feel, move, and care for ourselves. Alignment is not just a spiritual idea; it is a whole-body practice. When the body, mind, and spirit vibrate in harmony, we become clear channels for the flow of life. This section invites you to embody the truth you now carry, creating a coherence so strong that every cell, every thought, and every heartbeat becomes an act of alignment, devotion, ritual and faith.

Chapter 4:
The Body as a Temple: Healing Emotionally and Physically

Your body is more than a vessel—it is a sacred dwelling place, a living temple where the presence of the Divine meets the rhythms of the Earth. Every cell carries both memory and possibility, recording the imprints of your experiences while also holding the blueprint for perfect health. To honor the body is to recognize it as the house of the soul, designed not only to carry you through life but to communicate directly with the Source. Emotional wounds leave marks on this temple just as surely as physical injuries, and when left unhealed, they can dim the light within. But when we restore balance—releasing what no longer serves, nourishing what is life-giving—the temple begins to shine again. In this space, emotional release becomes physical healing, and caring for the body becomes an act of spiritual devotion.

Emotional Release and Physical Healing Connection

Emotions are not just passing moods; they are biochemical events that leave tangible traces in the body. Every unexpressed grief, unspoken truth, or restrained joy creates a physiological imprint—changing hormone levels, altering neural pathways, and influencing immune function. Over time, these stored emotions can manifest as chronic tension, fatigue, inflammation, and even illness.

This is not a flaw in your design—it is the body's way of holding what the mind is not yet ready to process. But what is unprocessed cannot stay hidden forever. The body will speak, first in whispers, then in shouts, until the message is heard. This is why moments of emotional release—whether through tears, laughter, prayer, breathwork, or creative expression—are so often followed by a sense of lightness, warmth, or unexpected physical relief.

Medical studies now confirm what ancient healers have always known: emotional states influence cellular repair, immune resilience, and even genetic expression. In the language of spirit, when you let go of what is heavy in the heart, you make room for the divine current to move freely through you. The body remembers how to heal because healing is its natural state.

Emotional release is not about dwelling in the past—it is about creating space for the body's innate intelligence to restore balance. When energy is freed, vitality returns, and the natural healing processes of the body accelerate. Many wisdom traditions describe the body as having energy centers—often called chakras—through which life force flows. When emotions are suppressed, these centers can become congested, much like a river blocked by debris. Over time, this blockage not only affects the flow within you but also your connection to the greater energy grid of the Earth. These blocks as "Emotional Molecules" can be passed through many generations, if not cleared. As we'll explore later, clearing and balancing these centers restores harmony both within your body and with the larger currents of creation.

Inflammation as Unprocessed Emotion

Inflammation is often spoken of as the root of disease—and while it can be triggered by diet, toxins, or injury, there is another source that is less visible but equally powerful: unprocessed emotion. When a strong emotion is suppressed rather than expressed, the body doesn't simply make it disappear. Instead, it stores the emotional "charge" in tissues, muscles, and even the fascia that wraps every organ and fiber of your being.

In this state, the body perceives an unresolved emotion as a lingering threat, keeping stress hormones elevated and immune responses on high alert. Over time, this constant low-level alarm feeds chronic inflammation, wearing down the systems designed to protect you. What begins as emotional tension can slowly translate into physical pain, autoimmune flare-ups, digestive issues, and other chronic conditions.

Ancient healing systems—from Chinese medicine to Ayurveda—have long understood this connection, mapping certain emotions to specific organs: grief to the lungs, anger to the liver, fear to the kidneys. Modern research in psychoneuroimmunology now supports these links, showing how emotional patterns influence immune function, hormonal balance, and the inflammatory cascade.

When we acknowledge and safely release stored emotions, we interrupt this cycle. The body can downshift from defense into repair, inflammation begins to subside, and healing accelerates. Seen in this light, emotional release is not just a psychological practice—it is an anti-inflammatory treatment at the deepest level of your being. And as your inner "fires" cool, the energy once consumed by constant defense becomes available again for vitality, joy, and creation.

Honoring the Body as the House of the Soul

Your body is more than a biological machine—it is the sacred dwelling place of your soul. Every beat of your heart, every breath you take, every movement you make is part of a living temple designed to hold the presence of the Divine. When you see your body in this way, caring for it is no longer about appearance or performance—it becomes an act of devotion.

This temple has its own language. It speaks through sensations, energy shifts, and subtle inner promptings. When the body is tired, it asks for rest; when it is depleted, it asks for nourishment; when it is stagnant, it calls for movement. To ignore these messages is to neglect the sanctuary that was entrusted to you at conception.

Honoring the body also means listening to its spiritual architecture. In many wisdom traditions, the body is understood to contain channels and centers through which life force flows—corridors of energy that connect the seen and unseen. Keeping these pathways clear through movement, breath, prayer, and mindful living allows the soul's light to fill every cell, strengthening both health and spiritual presence.

When you care for your body as the house of the soul, you are not indulging the physical—you are tending the altar of your being. And the more lovingly you tend that altar, the more radiant and resilient you become, both for your own journey and as a light for others.

The Role of Beauty, Sacred Space, and Feng Shui in Coherence

Beauty is not luxury—it is nourishment for the soul. A beautiful space invites the nervous system to relax, the mind to open, and the spirit to rise. Sacred spaces amplify this effect, creating an energetic container that supports peace, inspiration, and alignment. In these spaces, every detail matters: color, texture, light, placement. When each element harmonizes with the whole, the result is a frequency of coherence that the body and spirit instinctively recognize as home.

This principle is at the heart of **Feng Shui**, an ancient Chinese art and science of spatial alignment. Rooted in over 3,000 years of observation, Feng Shui, literally "wind-water") was developed to harmonize human life with the energies of the natural world. It is not random decoration—it is a precise method of measuring, orienting, and arranging spaces so that life force (qi) can flow freely and abundantly.

Historically, Feng Shui masters used the **luopan**, a specialized compass marked with concentric rings of calculations, to determine the optimal orientation of a home or sacred site. These calculations integrated astronomy, geography, mathematics, and the observation of seasonal cycles. A building's position relative to the sun and moon, the surrounding mountains and rivers, and the cardinal directions was considered essential to the well-being of its inhabitants.

This is the same universal intelligence expressed in sacred geometry: the ratios that govern the spirals of seashells and galaxies, the golden proportion that shapes both the human face and ancient temples, the alignments between celestial bodies like Earth, Mars, and the Sun that influence gravitational and electromagnetic fields. The human body itself reflects these harmonies—proportions from chin to forehead, navel to crown, echo the same mathematical relationships found in art, architecture, and planetary motion.

In Feng Shui, clutter is not simply untidiness—it is a physical blockage that mirrors energetic stagnation. A cluttered environment disrupts the natural flow of qi, much like a dam obstructs a river. By clearing and intentionally arranging our spaces, we remove obstacles to vitality, just as we release emotional blockages from the body. Both invite coherence.

Beauty and Feng Shui share the same purpose: to create environments that resonate with the frequency of harmony. When we live in such spaces, our energy naturally synchronizes with the greater order of life, supporting emotional balance, physical health, and spiritual clarity. In this way, your home, workplace, or any space you inhabit becomes not just a shelter, but a living partner in your awakening—a place where the God Spark can shine unobstructed.

Chapter 5: Food, Water, Fasting, and Detoxing

What we choose to put into—and withhold from—our bodies is not just a matter of health; it is a matter of frequency and light. Every molecule you ingest carries an energetic signature that either amplifies or diminishes the God Spark within you. In the modern world, it is easy to forget that food is not merely fuel, but information. It speaks to your cells, to your DNA, even to your consciousness, shaping how you think, feel, and heal. The same is true for water—the most abundant substance in your body and the most responsive to energy and intention. Science now confirms what spiritual teachers have said for millennia: the vibration of intentional prayer over food and drink literally alters their molecular structure, transforming what you consume into a carrier of blessing.

And just as food and water affect our frequency, so does the environment in which we share them. Gathering around a table with others—happy, laughing, talking, enjoying—infuses nourishment with joy. This is why the intentional table matters: fresh flowers, candles, a tablecloth, beautiful plates, and utensils. The visual beauty, the textures, the aromas, the colors—all of it speaks to the senses, inviting the body into harmony. Everything is everything is everything; no detail is too small when it comes to raising frequency.

I learned this in an unforgettable way on a healing trip to the site of the first atomic bomb test. After a long day, our group—which included several Japanese Zen monks and dignitaries—gathered in a small restaurant in Albuquerque, New Mexico, to celebrate. To my surprise, some of the monks ordered hot dogs, Frito pies, and beer. At the time, I was a strict vegetarian, not always the easiest dinner companion. Curious, I asked our translator why these pure Zen monks were ordering such foods. He laughed and said, "Oh Trish, they are not above the people they are with. They enjoy and appreciate the people and customs they are with. And besides, Dr. Trish, if your intention and frequency are right, you can transmute poison and it will be okay."

That moment changed me forever. I realized that love and presence with others are often more nourishing than the strictest diet. I stopped being so hard on myself with self-judgment about what I "should" or "should not" eat. I began to truly enjoy people as unique, different, and fascinating—allowing each shared meal to be a moment of connection and joy. And I saw that when I "let go and let God," or as my friend says of the Himalayan monks, *k'garnae*—it is all okay—my happiness rose, and so did the happiness of those around me.

Just as we are nourished by what we take in, we are renewed by what we release. Fasting and detoxing are ancient practices found in nearly every spiritual tradition—not as punishment or deprivation, but as pathways to clarity, cellular renewal, and deeper connection with the Source. Modern science now confirms what the mystics have always known: giving the body space to rest from constant digestion activates powerful repair mechanisms, allowing damaged cells to be recycled and inflammation to subside.

This chapter explores the sacred relationship between nourishment, cleansing, and frequency. We will look at how intentional fasting can reset the body's natural rhythms, how living water carries the memory of divine order, and how to choose foods that vibrate with the energy of life itself. Whether through the discipline of fasting, the blessing of pure water, the cleansing of toxins, or the joy of eating high-frequency foods, each choice becomes an act of alignment—feeding not only the body, but the light within it.

Fasting for Clarity and Cellular Renewal

Fasting is one of the oldest spiritual and physical practices in human history. Across cultures and religions—from ancient desert mystics to modern wellness seekers—it has been honored as a way to clear the mind, heal the body, and deepen communion with the Divine. When we step back from constant eating, we create a pause in which the body can redirect its energy away from digestion and into cellular repair, detoxification, and renewal.

Modern science now describes what sages have always known: fasting activates processes like autophagy, where damaged cells are broken down and recycled, and stem cell production, which supports regeneration. In energetic terms, fasting lightens the density of the body, allowing your frequency to rise and your inner light to become clearer and steadier. This is why so many report not only improved health but sharper intuition, heightened creativity, and a profound sense of spiritual connection when they fast intentionally.

Fasting can take many forms—intermittent daily windows, extended multi-day water fasts, or periodic cleansing days with broths and fresh juices. The key is intention. Just as prayer over food can change its molecular structure, entering a fast with prayer, gratitude, and clarity of purpose changes its energetic structure in you. You are not simply abstaining from food—you are aligning your body, mind, and spirit to receive light and life in a purer form.

In this way, fasting is both discipline and devotion. It honors the sacred temple of the body by giving it rest, and it honors the God Spark by allowing more of your inner radiance to shine. As with all practices in this chapter, the point is not punishment or rigid self-denial, but the joyful recognition that every choice you make—whether to eat, to drink, or to pause—can be an act of frequency alignment.

This very book was conceived in such a moment—waking in the middle of the night on the *fourth* day of what had begun as a three-day water-only fast. In the stillness and clarity of that night, the entire structure of *God Spark* poured into my awareness like a living transmission. It was as if my cells, unburdened and illuminated, became a clearer receiver for divine frequency. Each time I fast, I touch that same threshold—moments of profound spiritual experience where insight flows freely and the invisible becomes visible. Fasting, for me, is not just a practice; it is a doorway into the sacred.

Preparing for a Fast

Before beginning any extended fast, it's important to prepare your body and mind so the experience is one of choice and empowerment, not force. Start by simply practicing going without food for one meal. Notice the psychological programming that tells you it's "time" to eat. Realize that you have the power to choose differently, to act out of courage and self-discipline rather than habit.

For at least a week before your fast, remove sugar, soft drinks, and all highly processed, non-food items from your diet. This helps stabilize your blood sugar, reduce chemical cravings, and calm your nervous system—making the fasting period far easier. Eat whole, organic, and natural foods during this time.

The evening before your fast, enjoy a lighter-than-normal dinner made from fresh, natural ingredients, avoiding sugar entirely. This final meal will set the tone for the fast and make the transition smoother. After dinner, you will not eat again until the same time three days later, when you will break your fast.

During the fast, have Celtic, Himalayan, or sea salt available. A pinch dissolved in water replenishes trace minerals and helps curb cravings. A drop or two of pure lemon or peppermint Young Living Essential Oils in water (only if using true, food-grade oils with no additives) can add a pleasant taste and mild digestive support. Coffee or green tea are acceptable for many, though some use the fast as an opportunity to let go of caffeine entirely.

Breaking the Fast

Breaking a fast is as important as the fast itself. This is when your digestive system "wakes up" and needs to be treated with care:

1. **Start with a Liquid Nourishment** – Begin with a cup of warm, organic bone broth. Bone broth contains lysine and other amino acids that soothe the gut lining and prepare it to accept nutrients again.

2. **Wait 30–60 Minutes** – Give your system time to gently restart.

3. **Introduce Fermented Foods (Probiotics)** – Eat a small serving of organic sauerkraut, kimchi, or other fermented vegetables to reintroduce beneficial bacteria to the gut.

4. **Wait Another 30–60 Minutes** – Allow the probiotics to begin colonizing without competition from heavier foods.

5. **Add Prebiotic-Rich Plant Foods** – Create a salad with fresh greens, celery, parsley, mint, fresh herbs, seeds, raw nuts, avocado, and lightly steamed broccoli or cauliflower. Dress only with pure olive oil at this stage—no vinegar yet, as the gut may still be sensitive.

6. **Wait Another 30–60 Minutes** – Give your digestive system space to process each step fully.

7. **Finish with Clean Protein** – Choose an organic, clean source such as eggs, wild-caught salmon, grass-fed beef, or free-range, non-hormone chicken. Eat slowly and with gratitude.

When approached in this way, fasting becomes not an act of deprivation, but a sacred reset—a time when you honor your body's natural rhythms, cleanse your inner temple, and open your channel for deeper spiritual experience.

Structured Water and the Power of Intention

Water is the most abundant substance in your body, making up roughly 70–80% of your composition. Yet it is far more than a passive liquid—it is the most sophisticated information storage medium known in nature. Laboratory research has shown that water's capacity to store and transmit energetic information is at least **100 times greater than quartz crystals**—the same quartz we rely on for the precise memory and timing functions in computers, medical devices, satellites, and watches.

If quartz is trusted to store the data that runs our technology, consider the scale of influence water has inside you. Every cell, every organ, every drop of blood is bathed in water that is constantly recording and transmitting the energetic environment it encounters—your thoughts, your emotions, the words spoken to you, the music you listen to, the spaces you inhabit, and even the unspoken intentions of those around you. This stored information influences the way your body functions, how it heals, and the frequency you broadcast into the world.

The late Dr. Masaru Emoto brought this truth to life through his photographic studies of frozen water crystals. Water exposed to words of love, gratitude, and beauty formed luminous, symmetrical patterns—tiny masterpieces of sacred geometry. In contrast, water exposed to hatred, neglect, or discord produced fractured, chaotic patterns. His work visually confirmed what wisdom traditions have taught for millennia: water reflects the consciousness it encounters.

Physicist Dr. Bill Tiller expanded this understanding by demonstrating that focused human intention could alter the **pH of water dramatically** without adding any chemical substances—changing its chemistry purely through consciousness. His experiments provided hard evidence that human thought is not limited to the brain—it is a measurable, directive force that can reorganize matter at the molecular level.

This is why the words you speak and the thoughts you hold are not inconsequential. Words are carriers of frequency. The very term *spelling* reminds us that words are like spells—they create patterns, imprint memory, and shape reality. When you speak over water, food, or your own body, you are programming a living, responsive medium with your intention.

Blessing your water, praying over meals, or speaking affirmations into your heart are not symbolic gestures—they are acts of energy alignment. Structured water remembers, and because it flows through every part of you, its memory becomes your memory. When you infuse water with the frequency of love, gratitude, and truth, you are not only changing the water—you are changing yourself. Every sip becomes a sacrament, every drop a mirror of your highest intention, every molecule a carrier of coherence and light.

High-Frequency Food Compass for Vitality

The body is a living frequency instrument. Every bite you eat and every sip you drink either tunes it toward coherence or pushes it out of harmony. This is not opinion—it's measurable. Modern science confirms that all living matter emits a frequency, measured in megahertz (MHz). Healthy human cells vibrate between 62–72 MHz. Disease states are associated with much lower frequencies. The food you choose can either raise or lower your vibration immediately.

The Frequency Spectrum of Foods

- **High-frequency foods**: Fresh, organic fruits and vegetables, wild herbs, sprouts, raw nuts and seeds, naturally fermented foods, pure spring or structured water.
- **Moderate-frequency foods**: Lightly cooked vegetables, whole grains, wild-caught fish, organic eggs, clean plant oils (olive, coconut, flax).
- **Low-frequency foods**: Processed, refined, or heavily cooked foods; white sugar and flour; chemical additives; factory-farmed meats; rancid oils.
- **Zero or negative frequency foods**: Ultra-processed snacks, fast food, microwaved meals, foods with artificial colors, preservatives, or pesticides.

The Frequency of Illness and Health

Researchers in vibrational medicine have observed that the human body radiates measurable frequencies. When these frequencies drop, the body becomes more susceptible to illness. When they rise, vitality and coherence return.

Disclaimer: These numbers are offered for educational and inspirational purposes only. They are not medical advice and should not replace consultation with a qualified health professional.

State of Being	**Approximate Frequency (MHz)**
Death	25 MHz and below
Cancer	42 MHz
Candida/Yeast	55 MHz
Cold/Flu	57–60 MHz
Healthy Body (average)	62–78 MHz
Brain functioning optimally	72–90 MHz
Love, Gratitude, Prayer	90+ MHz

The Prayer Effect

Water and food store the memory of everything they encounter, as proven by Dr. Masaru Emoto's water crystal studies and Dr. Bill Tiller's pH experiments. Just as structured water responds to intention, food responds to blessing and gratitude. Saying a prayer over your meal—silently or aloud—literally changes its molecular structure, raising its frequency before it enters your body.

The Environment Multiplier

The frequency of your food is amplified or diminished by the environment in which you eat. Sharing a meal with people you love, laughing, and enjoying uplifting conversation increases your body's ability to absorb and integrate nutrients. Eating in a space with beauty—fresh flowers, candles, natural light, harmonious colors—further elevates the experience. This is why sacred meals in every tradition are shared with intention, community, and ceremony.

Your Compass in One Line

If it's alive, vibrant, and grown or raised with care—it will raise your frequency. If it's dead, processed, and stripped of its natural essence—it will lower it.

Simple High Frequency Practices

1. Bless your food and water before consuming.
2. Choose fresh, organic, seasonal ingredients when possible.
3. Eat with others in an environment of joy and beauty.
4. Limit or avoid chemical additives, refined sugar, and processed oils.
5. Hydrate with pure, structured, or mineral-rich water throughout the day.

When you combine high-frequency foods, conscious blessing, and intentional environment, you create more than a meal—you create a vibrational upgrade for your body, mind, and spirit.

You were never meant to live disconnected from the intelligence of what you eat and drink. This wisdom—that food is information, that water is memory, that fasting is renewal—has been known in every ancient culture and only recently forgotten in modern life. Science has now confirmed what the elders and mystics always taught: intention changes matter, and the frequency of what you consume becomes the frequency of who you are.

This is why every choice matters. A sip of blessed water. A bite of living food. A fast entered with reverence. A table set in beauty and shared in joy. These are not luxuries—they are acts of remembering the original design, where nourishment was sacred and the body was honored as the temple of the soul.

If this was taught to you from childhood, you would never again see eating or drinking as mundane. You would know that every meal is an opportunity to raise your frequency, to strengthen the God Spark within, and to infuse your very cells with light. And you would understand, with certainty, that what you choose today will shape not only your health, but the clarity of your spirit for all your days to come.

Chapter 6: Sacred Movement, Color, and Joy: Activating Vibrancy

Once the body is nourished, cleansed, and aligned, the next step is to awaken its natural vibrancy. Energy thrives on motion, beauty, and delight. Movement, color, and joy are not decorative extras—they are ancient, essential medicines for the body, mind, and spirit. Every movement you make, every color that surrounds you, every note of music you invite into your space sends a signal through your nervous system, shifts your chemistry, and alters the frequency you radiate. These signals are powerful enough to influence mood, raise immune function, and even bring the heart and brain into greater coherence.

From the flowing grace of tai chi to the exuberant freedom of dance, from walking barefoot on the earth to surrounding yourself with colors that nourish your soul, these practices open dormant energy pathways and reconnect you to the joy of being alive. Joy is not merely an emotion—it is a measurable frequency that reorders your entire system toward health, resilience, and expansion. Music, too, is vibration made audible, capable of instantly lifting your spirit or wrapping you in deep peace.

In this chapter, we will explore movement as medicine, color as frequency therapy, the neurochemical power of joy, and the way music can become a daily tool for vibrational elevation. Along the way, we will look at the ancient and scientific role of rhythm, cycles, and sacred numbers—from the 108 beads of a prayer mala to the resonant tones of bells and chimes—showing that pattern and harmony are woven into the design of life itself. By the end, you will have simple, joyful practices that infuse your days with vitality and activate the God Spark in every cell.

Movement as Medicine

The body is designed to move. Every joint, muscle, and organ system thrives on circulation, oxygenation, and the rhythmic flow of energy. Yet in modern life, many people spend the majority of their hours sitting—draining vitality, slowing metabolism, and causing the subtle energy channels of the body to stagnate. Movement is not only for fitness; it is a primary way to keep your life force, or qi, circulating freely.

When you move, you stimulate lymphatic flow, allowing the body to detoxify naturally. Your heart rate rises, sending oxygen-rich blood to every cell. Endorphins—your body's natural mood elevators—are released, lifting your spirit and lowering stress. On an energetic level, movement clears stagnant frequencies and brings your field back into coherence. Science confirms this: rhythmic movement has been shown to synchronize brain hemispheres, improve emotional regulation, and even enhance cognitive performance.

Across cultures, movement has always been used as medicine. In China, the flowing postures of **tai chi** and **qigong** were developed not only for balance and strength, but to align the body's meridians for optimal energy flow. In India, **yoga** emerged as a way to prepare the body for deep meditation, keeping the spine supple and the organs toned so prana—the life force—could move without obstruction. Indigenous peoples worldwide have used **dance** as a way to celebrate, to heal, and to unite community energy. Even the simplest form of movement—**walking barefoot on the earth**—has profound effects, allowing your body to discharge built-up static energy and absorb the Earth's stabilizing electromagnetic frequency.

Movement also serves as an emotional release. Trauma, grief, and unspoken emotions can lodge in the muscles and fascia, creating tightness or chronic discomfort. Gentle stretching, shaking, or dance can help the body "speak" what the mind has silenced. Many find that a slow walk, a flowing yoga sequence, or even swaying to music unlocks tears, laughter, or deep peace without the need for words.

To make movement a daily medicine, it doesn't have to be long or complicated. A five-minute barefoot walk, a few yoga postures, dancing in your living room, or practicing tai chi in the park can reset your nervous system, lift your mood, and raise your frequency. The key is to move with awareness and intention—knowing that each step, stretch, or turn is tuning your body like an instrument.

Sacred Numbers, Cycles, and Rhythms

Numbers are more than tools for counting—they are the language of pattern, proportion, and cosmic order. Across cultures and throughout history, certain numbers appear repeatedly in the architecture of sacred sites, the design of the human body, and the cycles of the Earth and heavens. These numbers were embedded into temples, rituals, music, and movement because they were understood to carry a frequency that could restore balance and awaken consciousness.

One of the most universal is **108**. In Hinduism and Buddhism, prayer malas contain 108 beads, each bead marking one repetition of a mantra to align the mind and heart. Many temples are built with 108 steps leading from the entrance to the inner sanctum, and bells in shrines are often struck exactly 108 times to complete a sacred cycle. This number is not arbitrary. Ancient Vedic teachings describe **108 energy lines (nadis)** converging at the heart chakra—the spiritual center of love, compassion, and divine connection. In astronomy, the average distance from Earth to the Sun is approximately **108 times the Sun's diameter**, and the distance from Earth to the Moon is about **108 times the Moon's diameter**. This same number is echoed in the Catholic tradition, where certain rosaries contain 108 beads, and the ringing of three bells in succession is used to mark sacred moments, symbolizing the Trinity.

The great inventor **Nikola Tesla** believed that the numbers **3, 6, and 9** held the key to unlocking the secrets of the universe. He often performed actions in multiples of three, convinced these numbers reflected a universal energy pattern. Ancient builders also understood this. Many pyramids, cathedrals, and megalithic sites incorporate these ratios in their steps, chamber dimensions, and alignments with celestial bodies.

Another number with deep symbolic resonance is **4**. It represents foundation and stability: the four directions (north, south, east, west), the four elements (earth, air, fire, water), the four seasons, and the four phases of the moon. In sacred movement and ceremonial dance, circles are often divided into four quadrants, each honoring one aspect of the natural order.

Numbers also carry deeply personal meaning. Repeating sequences such as **333**, **1111**, or **777** are often experienced as moments of divine alignment or messages from the unseen. In moments of grief, transition, or awakening, these patterns can serve as reassuring markers that we are in the flow of something far greater than ourselves.

Science now affirms that rhythm and repetition are not only symbolic—they have measurable effects on the body and mind. Neuroscience shows that repetitive patterns in movement, sound, or vision can shift brainwave activity into more coherent states, enhancing intuition, focus, and emotional regulation. This is why prayer cycles, mantras, drumbeats, walking meditations, and bell ringing are so common in spiritual traditions. They entrain the body, mind, and spirit to a steady frequency of order and harmony.

From the measured steps of Buddhist walking meditation to the chiming of temple and church bells, from the counting of rosary prayers to the beating of ceremonial drums, sacred numbers, cycles, and rhythms remind us that we live in a mathematically ordered creation. They are not superstition, but reflections of the same harmony that governs the orbit of planets, the beating of our hearts, the rhythm of our breath, and the steps we take in mindful movement. When we align ourselves with these patterns, we step into resonance with the very architecture of the universe.

Sacred Numbers: Science and Spirit

Number	Scientific / Natural Connection	Spiritual & Cultural Connection	Meaning / Frequency
3	Triads appear in atomic structure (proton, neutron, electron) and geometric stability (triangle is strongest shape).	Holy Trinity in Christianity, Triple Gem in Buddhism, Triple Goddess in Celtic tradition. Church bells often rung in threes.	Harmony of body–mind–spirit; divine wholeness.
4	Four directions, four seasons, four phases of the moon; genetic code built from four nucleotides (A, T, G, C).	Four Gospels, Native American Medicine Wheel, four Hindu Vedas.	Stability, foundation, grounding.

Number	Scientific / Natural Connection	Spiritual & Cultural Connection	Meaning / Frequency
6	Hexagonal geometry found in snowflakes, honeycombs, and crystal structures.	Six directions in sacred space (N, S, E, W, above, below); six days of creation.	Balance, order, creative structure.
7	Seven colors of visible light, seven notes in the musical scale, seven main chakras.	Seven days of the week, Biblical symbolism of completion, seven heavens in Islam.	Spiritual awakening, completeness, divine cycles.
9	Multiples of three, seen as "completion" number in mathematics; nine-month human gestation.	Nine choirs of angels, Norse Nine Worlds, Egyptian Ennead.	Fulfillment, mastery, higher consciousness.
12	Twelve lunar months in a solar year; human body has 12 paired cranial nerves.	Twelve apostles, Twelve tribes of Israel, Twelve signs of the zodiac.	Cosmic order, governance, spiritual authority.
40	Forty weeks of pregnancy; biblical "testing" cycles often linked to purification.	Moses' 40 days on Mount Sinai, Jesus' 40 days fasting, Islamic 40-day mourning period.	Transformation, preparation, spiritual renewal.
108	Earth–Sun distance is ~108× Sun's diameter; Earth–Moon distance is ~108× Moon's diameter; human heart has 108 energy lines (nadis).	108 beads in malas and certain rosaries; 108 steps in temples; bells rung 108 times.	Cosmic connection, heart alignment, universal harmony.
333	Repeating triple amplifies the vibrational meaning of "3"; seen in harmonic ratios in sound.	Often experienced as an "awakening number" or sign of divine presence; associated with angelic communication.	Support, encouragement, divine companionship

These numbers are not superstition—they are part of the mathematical and vibrational architecture of life itself. They appear in the orbits of planets, the petals of flowers, the proportions of the human body, the cycles of the seasons, and in the sacred architecture of temples, cathedrals, pyramids, and monuments across the world. Movement, prayer, and sound rituals based on these numbers are found in every major tradition because they entrain the body and mind into harmony with the greater whole. When you walk, breathe, chant, or even ring a bell in sync with these patterns, you are aligning yourself with the same cosmic order that keeps the stars in place, the tides in rhythm, and the heart beating in perfect time.

Color as Frequency Therapy

Color is more than decoration—it is visible vibration. Each color corresponds to a specific wavelength of light, measured in nanometers, and each wavelength carries a frequency that interacts with the body and mind. Just as sacred numbers and rhythms can entrain your energy, so can color. Light enters through the eyes and skin, triggering chemical reactions in the brain and nervous system, influencing mood, hormone production, and even cellular repair.

In the electromagnetic spectrum, colors move from slow, long wavelengths (reds and oranges) to fast, short wavelengths (blues and violets). Ancient healers and modern researchers alike have observed that these frequencies correspond with different effects on the human system:

- **Red**: Stimulating, warming, and energizing; linked to increased circulation and vitality.
- **Orange**: Uplifting and creative; often used to spark enthusiasm and social connection.
- **Yellow**: Clarifying and focusing; stimulates the mind and promotes optimism.
- **Green**: Calming and balancing; resonates with the heart and supports emotional harmony.

- **Blue**: Cooling and soothing; reduces stress and slows an overactive mind.
- **Violet**: Inspiring and spiritual; associated with meditation, intuition, and higher states of consciousness.

Many ancient traditions understood the relationship between color, light, and life force. Egyptian temples contained chambers painted with specific hues to create different healing effects. Tibetan monks use brilliantly colored mandalas to focus meditation, while Christian cathedrals filter sunlight through stained glass to bathe worshippers in symbolic colors of devotion. Even modern hospitals use color strategically—soft greens and blues in recovery rooms to promote calm, warmer tones in therapy spaces to encourage interaction.

Color therapy is not only visual—it is also energetic. The chakras, or energy centers of the body, are often associated with specific colors, each reflecting a frequency that supports that center's function. Surrounding yourself with the colors that match your physical or emotional needs can help bring you back into balance. This can be as simple as wearing clothing in a certain hue, eating richly colored fruits and vegetables, or decorating your home with intentional color placement.

In essence, color is light made visible, and light is one of the purest forms of frequency. When you choose colors consciously—whether in your clothing, your surroundings, or your food—you are not just making an aesthetic choice. You are tuning your personal field, much like a musician tunes an instrument, so that your energy resonates with the qualities you wish to embody.

Clothing the Body in Frequency: The Hidden Power of Fabrics

Every inch of our skin is a living receptor, a sensory and energetic interface between our inner world and the world around us. Science confirms that what touches our skin can be absorbed into our bloodstream within minutes. This is not just true for lotions and oils—it is also true for the fabrics we wear and sleep in.

Natural fibers such as **linen, silk, organic cotton, and wool** carry their own measurable frequencies, and when we are wrapped in them, those vibrations interact directly with our body's electromagnetic field. Studies and frequency measurements have shown that:

- **Linen** can measure around **5,000+ MHz**, making it one of the highest-frequency fabrics known. Linen is naturally antibacterial, cooling, and long-lasting—often improving with age.
- **Silk** resonates in a high, smooth frequency range that feels both luxurious and calming to the nervous system. It has been used for centuries as a healing and protective fabric.
- **Organic cotton** carries a frequency closer to the natural rhythm of the Earth itself, grounding and stabilizing the wearer.
- **Wool** is warming and protective, with insulating properties that shield against environmental stressors.

When fabrics are **organic and chemical-free**, they maintain their full vibrational signature. However, when they are treated with pesticides, chemical dyes, or synthetic softeners, their frequency drops dramatically, sometimes into ranges that drain the body's natural energy.

Synthetics like polyester, acrylic, nylon, and fabrics blended with plastics hold **low, static-like frequencies**. Over time, wearing these materials—especially against the skin—can subtly weaken the body's own frequency field, making it more vulnerable to stress and fatigue.

It is not just about what we wear during the day. **Nighttime** is when our bodies enter deep regeneration, repair, and energetic recalibration. Sleeping in organic, high-frequency fabrics—especially **linen, cotton, or silk bedding and sleepwear**—allows the body to rest in harmony with supportive, life-enhancing vibrations. Imagine the difference between lying on breathable, organic linen sheets washed in natural soap, versus synthetic microfiber washed in chemical detergent. One allows your cells to breathe; the other forces them to adapt to a low-grade chemical and frequency assault.

Your laundry choices matter, too. Washing clothes and bedding in organic, plant-based soaps preserves their frequency and prevents chemical residues from entering the skin. Even the scent matters—natural essential oils not only add fragrance but also elevate the vibrational field of the fabric itself.

True luxury is not about a price tag. It is about **the purity, integrity, and frequency** of what you surround yourself with every day. A $30 organic linen pillowcase can be infinitely more healing and luxurious than a $300 synthetic designer one. Your body knows the difference. It responds with ease, comfort, and often a subtle joy—like a sigh of relief from the soul.

What you wear and sleep in is not just a matter of style. It is a form of **energetic nutrition**, feeding your body's frequency in every moment.

Fabric Frequency Chart

(Approximate MHz Vibrations – Higher is More Life-Supporting)

Fabric Type	Approx. Frequency (MHz)	Notes & Benefits
Linen (organic)	~5,000+ MHz	Highest known natural fabric frequency; antibacterial, cooling, strengthens immune field
Silk (pure, untreated)	~3,000–4,000 MHz	Luxurious, calming to nervous system, protective
Wool (organic)	~3,000 MHz	Warming, protective, shields from environmental stress
Organic Cotton	~1,000–1,500 MHz	Grounding, breathable, resonates with Earth's field
Hemp (organic)	~1,500–2,000 MHz	Durable, antimicrobial, balances energy
Cashmere (pure)	~2,500–3,000 MHz	Soft, insulating, high comfort frequency
Synthetic Blends	~0–100 MHz	Polyester, nylon, acrylic; low or static frequency, can drain vitality
Plastics / Vinyl	0 MHz	No measurable life-supporting frequency

Tips for Daily Life

- Always wash new fabrics before wearing—preferably in **plant-based, chemical-free soap**.
- Infuse your laundry water with a drop of **pure essential oil** (e.g., lavender, lemon, rose) to raise frequency.
- Avoid synthetic dryer sheets—they coat fabrics with chemicals that lower their frequency.

How Fabric Frequencies Are Measured

The frequency values for fabrics are not guesses—they come from methods similar to those used in testing essential oils, foods, and even human energy fields.

Researchers and energy medicine practitioners have used:

- **Frequency Counters & Spectrum Analyzers** – Tools adapted from radio-wave and engineering labs, calibrated to measure subtle electromagnetic resonance.
- **Tesla Coil Resonance Testing** – Determines how a fabric holds and transmits energy by exposing it to a controlled EM field.
- **Gas Discharge Visualization (GDV)** – A photographic method that records the corona discharge around a material to detect its energetic field.
- **Applied Kinesiology & Biofeedback Devices** – While more indirect, these test how the body's strength or stress response changes when in contact with different fabrics.
- **Electro-Photon Imaging (EPI)** – Used in scientific studies to see how materials influence biofields and biophoton emissions.

Key Findings

- Natural fibers like **linen** and **silk** consistently show **thousands of MHz higher** than synthetics.
- Synthetics not only fail to contribute to biofield vitality—they often act as energy **drains**, pulling measurable life force away from the body.

- When fabrics are treated with **chemical dyes, flame retardants, or softening agents**, their frequency drops sharply—even in organic fibers.

Why It Matters

Your skin is your largest organ, and it's conductive—meaning it acts like an antenna for the frequencies around you. What you wear all day, and what you sleep in at night, is in constant contact with your nervous system, lymph, and bloodstream. Choosing **pure, untreated, natural fabrics** is like choosing food for your skin: it nourishes or depletes you every single day.

Dance, Joy Practices, and the Dopamine Effect

Joy is not a frivolous indulgence—it is a biochemical and spiritual necessity. When you move your body in ways that feel good, whether it's dancing alone in your kitchen, swaying to live music, or joining a community drum circle, you trigger the release of dopamine, endorphins, and oxytocin—nature's own "happiness cocktail." These neurochemicals are not only mood elevators; they are immune boosters, anti-inflammatory agents, and frequency amplifiers.

Dance is one of the oldest forms of healing known to humanity. From the sacred temple dances of India, to African tribal celebrations, to the whirling of the Sufi dervishes, movement has always been a bridge between the human and the divine. When you dance, you invite energy to flow freely through the body's meridians, clearing stagnation and awakening a vibrancy that radiates into every cell.

Joy practices—laughter, singing, play, and lighthearted interaction—work in much the same way. They break the grip of stress, shift the brain into creative problem-solving modes, and help the body remember its natural state of balance. Something as simple as five minutes of free dance or genuine laughter each day can rewire your neural pathways toward optimism and resilience.

The key is presence. Dance not to necessarily "perform," but to inhabit your body fully, letting music and movement dissolve self-consciousness. Joy is the highest frequency you can generate without effort, and it acts like a magnet for synchronicity, clarity, and connection.

Using Music to Boost Frequency

If movement is medicine for the body, then music is medicine for the soul. Every note carries a frequency, and these frequencies influence the rhythm of your heartbeat, the coherence of your brainwaves, and even the crystalline structure of the water within your cells.

Scientific studies and ancient traditions agree: music is not just entertainment—it is a tuning fork for the human spirit. Certain tones and harmonics have measurable effects on the nervous system, immune function, and emotional states:

- **432 Hz** – Known as the "natural frequency of the universe," 432 Hz is mathematically aligned with the golden ratio and the rhythms of nature. Many listeners report it feels more calming, heart-opening, and spiritually grounding than the standard 440 Hz used in most modern music.

- **528 Hz** – The "Love Frequency" or "Miracle Tone," linked to DNA repair, cellular regeneration, and feelings of unconditional love.

- **639 Hz** – Resonates with the energy of relationships, connection, and harmony, often used in sound healing to mend emotional rifts.

- **741 Hz** – Cleansing frequency that assists in detoxifying cells and promoting mental clarity.
- **852 Hz** – Associated with awakening intuition and realigning with your spiritual path.

Ancient cultures intuitively understood this. Gregorian chants, Tibetan singing bowls, Native American flutes, and Indian ragas are all examples of music intentionally tuned to affect the mind, body, and spirit. Each tradition used specific scales, rhythms, and instruments to evoke states of peace, joy, courage, or divine connection.

When you listen with intention, music becomes an energetic tool. This is why I often recommend creating **personal frequency playlists**—collections of songs or soundscapes that shift your state in predictable, positive ways. Your playlists might include:

- Classical compositions tuned to 432 Hz for deep focus or meditation.
- Nature sound recordings layered with healing tones for relaxation.
- Uplifting world music for joyful movement and celebration.
- Sacred chants or mantras for spiritual alignment.

Streaming platforms like Spotify or Apple Music have entire collections dedicated to healing frequencies, binaural beats, and Solfeggio tones. Play them in the background while you work, during meditation, or as a sonic "energy bath" while you rest.

The key is to engage with music as you would a trusted healing ally—choosing it consciously, receiving it fully, and noticing how your body and emotions respond. Over time, you will begin to feel how sound can realign your frequency as powerfully as food, movement, or breath.

Sound and Color Synergy: The Frequency of Joy

Sound and color are two of the most immediate ways to shift the human frequency. Ancient cultures understood that certain tones and hues could alter emotional states, heal the body, and elevate consciousness. Today, science confirms what our ancestors practiced instinctively — that specific sound frequencies and color wavelengths resonate with the body's cells, nervous system, and energy field in measurable ways.

As many times as I repeat it, I can't emphasize enough, how effective healing sound frequencies are and how much they can impact your life; such as **432 Hz** (associated with harmony and natural resonance), **528 Hz** (often called the "Love" or DNA repair frequency), and **639 Hz** (promoting connection and empathy) work subtly yet powerfully to restore coherence in the body. Gregorian chants, crystal singing bowls, classical ragas, and nature sounds each carry vibrational signatures that influence brainwave patterns, heart rate, and emotional equilibrium. By intentionally surrounding yourself with music tuned to these healing frequencies, you invite your body into a state of harmony and receptivity.

Color works in the same way. **Warm tones** like gold, coral, and crimson stimulate and energize, while **cool shades** like teal, lavender, and silver soothe and balance. Sunlight — nature's full-spectrum therapy — is one of the most potent healers, offering every color frequency in perfect proportion. Wearing colors intentionally, decorating with harmonizing palettes, and seeking out natural light daily are not merely aesthetic choices — they are acts of energetic calibration.

But there is another often-overlooked factor in how we experience sound, color, and joy: **the materials we place directly on our skin**. Your clothing and bedding are in constant contact with your largest organ — the skin — which transmits information directly to your nervous system and bloodstream.

Natural fibers like **linen, silk, wool, hemp, and organic cotton** carry measurable high frequencies because they come from living sources. Linen, for example, has been measured at a frequency around **5,000 Hz**, far higher than synthetic fabrics, which can register close to zero. Silk soothes the nervous system, cotton allows the skin to breathe, and hemp offers antimicrobial properties — all while resonating at frequencies that support health and vitality.

The opposite is true for synthetic fabrics like polyester, nylon, or acrylic, which not only hold lower energetic signatures but can create static buildup, block natural energy flow, and, when treated with chemical finishes, introduce toxins into your body through prolonged skin contact. Over time, these subtle influences affect mood, sleep quality, and even the body's electromagnetic field.

When you combine sound and color therapy with **high-frequency fabrics**, you create a fully immersive vibrational experience. Imagine moving through your day wearing breathable, living fabrics in colors that support your emotional state, while listening to music tuned to healing frequencies. Or sleeping in organic linen sheets — the gentle texture against your skin — as a soft frequency playlist fills your room and moonlight filters through gauzy curtains.

This is joy as frequency in action — an intentional orchestration of sensory inputs, each one chosen for its ability to lift you higher. Movement, music, color, and the sacred intimacy of touch through fabric all converge to create a lived experience of vibrancy.

Daily Joy Frequency Ritual

A simple 10-minute practice to elevate your vibration every day.

1. Set Your Intention (1 minute)

- Before you start, pause and choose a feeling or frequency you wish to embody today — joy, peace, confidence, love.
- Say it aloud: *"I choose to carry the frequency of [chosen word] today."*

2. Dress in High Frequency Fabrics (2 minutes)

- Select clothing made of **linen, silk, hemp, or organic cotton** in a color that supports your mood.
  - **Need calm?** Choose blue, lavender, or soft green.
  - **Need energy?** Wear gold, coral, or vibrant red.

3. Activate with Sound (3 minutes)

- Play music tuned to a healing frequency that matches your intention:
  - **432 Hz** – Harmony and grounding
  - **528 Hz** – Love and cellular healing
  - **639 Hz** – Connection and empathy
- Let the music wash over you — notice your breath deepen and your posture open.

4. Move with Joy (2 minutes)

- Dance freely, sway, stretch, or walk barefoot on natural ground.
- Imagine each movement "painting" your energy field with color and sound.

5. Hydrate with Intention (2 minutes)

- Pour a glass of water and speak your intention into it.
- Hold the glass for a moment, visualizing the water storing your chosen frequency, then drink slowly.

Optional: Repeat this ritual before bed using softer colors, calming music (like 528 Hz or Gregorian chant), and organic bedding to create a peaceful night's frequency.

Scan this code to enter the Healing Frequency Playlist I use for myself from my personal Spotify Playlist as I share with you. I hope you like it.

Chapter 7: Sacred Sexuality and Life Force: Honoring the Divine Energy of Creation

Sacred sexuality is one of the most profound and misunderstood gifts we have been given by God. Across centuries and cultures, this life-giving force has been revered not only as the power to conceive new life, but as a sacred pathway to divine remembrance. Whether expressed alone in deep prayer and meditation, or shared in conscious union with a beloved partner, this energy carries within it the potential to awaken, heal, and elevate us to our highest self.

Modern science is only beginning to catch up with what the mystics and seers of old have long known. At the moment of conception, researchers have documented a visible spark—a literal burst of light—triggered by the union of sperm and egg. This zinc-fueled photonic flash is more than cellular ignition; it is a microcosmic reflection of the great ignition of creation itself. Just as the universe was birthed from a radiant explosion of energy, so too are we born of light.

But this spark is not limited to the beginning of physical life. Our bodies remain alight with electromagnetic potential. Every heartbeat, every breath, every touch carries within it an energetic signature. Studies by institutions like HeartMath have shown that during deep emotional or spiritual connection, the heart rhythms of two individuals can synchronize, and brainwave patterns can harmonize—offering measurable evidence that sacred union is not just symbolic; it is scientifically real.

In Eastern traditions, this creative current is known as Kundalini—a coiled life force energy resting at the base of the spine. When awakened, it rises like a golden river of light, activating the body's energetic centers and purifying the nervous system. In Christian mysticism, similar states are described during moments of divine ecstasy and deep contemplative prayer. Saints such as Teresa of Ávila and John of the Cross spoke of fire, light, and all-consuming union with God. In Catholic tradition, we still see traces of this sacred technology: the anointing with oils, the ringing of bells in threes, the lighting of candles—all subtle acts of vibrational alignment.

Whether awakened through Tantra's breath and movement practices, the stillness of Christian contemplation, or the intimate, reverent exchange between two lovers, this force is holy. It is a spark of God's own essence within us, placed not to shame or suppress, but to guide us into full embodiment of our divine inheritance. This energy, when approached with love, reverence, and conscious intent, becomes the bridge between Heaven and Earth.

You were never meant to feel powerless in your own body. This is your birthright: to be a keeper of light, a sacred vessel of creation, and an active participant in the divine unfolding of life.

Kundalini and Sexual Energy as Divine Creative Force

At the heart of sacred sexuality lies the truth that our sexual energy is not merely hormonal or physical—it is a **direct current to the Divine**. Ancient traditions from around the world have spoken of this power, often hidden beneath layers of metaphor, parable, or sacred secrecy. Kundalini, often described as a serpent of light coiled at the base of the spine, rises when we live in alignment with love, devotion, and truth. It spirals upward through the central channel of the body, activating chakras, opening awareness, and transforming our perception of what it means to be alive.

This rising is not metaphorical alone. It is deeply physiological. Those who have experienced it describe sensations of heat, light, bliss, involuntary movements, or the feeling of being washed by an inner flame. Scientific studies on altered states of consciousness confirm that in moments of deep ecstasy or spiritual union, the body enters measurable states of coherence. Brain waves synchronize. Hormones linked with healing and euphoria—like oxytocin, serotonin, and dopamine—are released in surges. The electromagnetic field of the heart expands.

In the Zero Point Field—a term from quantum physics describing the still-point of potential energy beneath all creation—energy naturally moves toward perfect harmony. When our sexual energy is activated in alignment with sacred love, we become instruments of this universal coherence. We don't just feel different; we become radiant.

The spark of conception mirrors this truth. The union of sperm and egg is a moment of perfect alignment, a gateway between realms. The burst of light that accompanies it is a symbol and a signal—an echo of divine intention. Some teachings suggest that during sacred union, whether through intimacy with a partner or deep connection with God, similar bursts of light ripple through the body's energetic field like stars aligning. This is not fantasy. Early research into biophoton emissions confirms that the body emits light, and that this light increases during moments of joy, meditation, prayer, and loving connection.

This is why many ancient traditions protected sexual energy with such care—not out of repression, but reverence. The force that creates galaxies flows through your body. To direct that force with consciousness is to step into your full power as a spiritual being in form. When sacred sexual energy is awakened and guided with reverence, it becomes a healing current—capable of regenerating the body, illuminating the mind, and deepening our communion with the Divine.

Modern mystics and neuroscientists like Dr. Joe Dispenza have shown that in advanced states of meditation, individuals produce high gamma brain waves—associated with peak states of awareness and insight—while simultaneously entering heart-brain coherence. Participants often describe sensations of flooding light, divine presence, and the dissolution of all separation. Instruments have even recorded subtle bursts of biophotonic light—tiny flashes radiating from within the cells—during these mystical states.

The Sacred Body Knows. It Remembers. And it Responds.

This rising of energy, this inner alignment, this communion with Spirit is not reserved for saints, monks, or mystics. It is available to all. Whether in sacred intimacy or solitary prayer, the Divine lives in your breath, your cells, your heart, your longing.

The mystics of Christianity, though they may not have called it Kundalini, described the same sacred phenomenon. Saint Teresa's "interior castle" and her descriptions of spiritual rapture mirror the experiences of tantric awakening. The anointing of the body with sacred oils, the use of breath and posture in prayer, and the prostration before the altar—these are all **echoes of** practices found in other spiritual systems, all pointing toward the same universal truth: **the body is a holy vessel**, and the life force that flows through it is sacred.

When we remember that the same current that creates stars pulses through our own nervous system, shame evaporates. Fear dissolves. We are restored to reverence.

Every breath becomes an act of holy creation.

Healing Sexuality Sacred Practice

A 5-Minute Heart–Soul Alignment Ritual

1. **Posture of Honor** – Stand or sit tall. Gently squeeze your shoulder blades together so your chest opens and your heart rises. Keep your head aligned as though wearing a crown. This physical openness signals safety to your nervous system and allows the heart to beat freely.

2. **Breath of Release** – Inhale deeply through the nose for a count of 6, imagining divine light filling your body from the base of the spine to the crown. Exhale slowly for a count of 8, visualizing any shame, fear, or old imprints dissolving into the earth.

3. **Root Reconnection** – Place one hand over your lower abdomen and the other over your heart. Silently affirm:
"I am safe in my body. I am worthy of joy. I am free."

4. **Light Transmission** – Visualize a golden current rising from the earth through your feet, filling your lower three chakras,

softening the heart, and flowing up through the throat, crown, and beyond. See yourself held in divine presence.

5. **Closing Seal** – Rest both hands over your heart. Whisper a prayer or scripture that affirms your sacred worth. Hold stillness for at least one full minute, letting the feeling anchor deep into your cells.

Practices for Harmonizing Masculine & Feminine Energies

Every human being carries both masculine and feminine energies—regardless of gender.

- **Masculine energy** expresses as structure, action, focus, protection, and clear direction.
- **Feminine energy** expresses as receptivity, intuition, nurturing, creativity, and flow.

When these energies are balanced, we feel whole, clear, and powerful. When they are out of balance, we may feel ungrounded, indecisive, overly rigid, emotionally shut down—or confused about our sexual drive and ability to give or receive pleasure.

The Biological Mirror

From a medical perspective, both men and women produce estrogen, progesterone, and testosterone—just in different ratios. These hormones influence not only mood, energy, and clarity of thought, but also **libido, sexual performance, and the capacity to experience pleasure.**

When hormones are in harmony, sexual energy flows naturally—heightening intimacy, creativity, and vitality. But when they are disrupted, sexual desire can vanish or become distorted, leading to shame, frustration, or a sense of disconnection from one's own body.

Modern life is filled with influences that **interfere with natural hormone balance** and distort sexual energy:

- **Processed foods & excess sugar** – Disrupt insulin and cortisol, indirectly affecting sex hormones.
- **Plastics & chemicals (BPA, phthalates)** – Mimic or block estrogen and testosterone, dulling sexual vitality.
- **Pharmaceutical side effects** – Some medications alter hormone production or sensitivity, impacting arousal.
- **Fluoride in water & toothpaste** – Contributes to pineal gland calcification, dulling intuition and hormonal regulation via the brain's endocrine centers.
- **Chronic stress** – Floods the body with cortisol, suppressing reproductive hormones and lowering libido.

The Pleasure Circuit Crisis

Beyond physical toxins, **modern society is addicted to overstimulation**—and it's blowing the fuses in our natural pleasure circuits.

- **Technology & social media** flood the brain with dopamine hits from constant scrolling, likes, and novelty.
- **Artificial environments** keep us indoors, depriving us of sunlight and grounding, which are essential for hormonal balance and circadian rhythm regulation.

- **Pornography** can short-circuit the brain's reward system, creating unrealistic expectations of arousal and intimacy, often numbing the ability to respond to real-life connection.

- **Excess stimulation in all areas**—loud entertainment, high-sugar diets, non-stop notifications—can overload the nervous system much like a circuit board under too much current. Eventually, the system shuts down or becomes desensitized.

True pleasure—whether sexual, creative, or emotional—requires sensitivity, not overstimulation. If every circuit is overloaded daily, the nervous system loses the ability to feel the subtle, nourishing currents of real intimacy and connection.

Sexual Wholeness

Balancing masculine and feminine is **not** about becoming genderless—it is about reclaiming the **full spectrum** of our human potential, including healthy sexual expression. A man can stand fully in his grounded strength while embracing gentleness and emotional depth. A woman can radiate intuitive softness while embodying fierce clarity and sovereign power.

Sexual pleasure is not simply a side effect of attraction—it is a natural byproduct of life force flowing freely through the body. When we embrace both polarities within, we create safety, openness, and vitality that allow sexual energy to circulate without shame or force. This is the sacred marriage of polarities—what Taoists call the union of yin and yang.

Daily Practices for Harmony

1. **Nutrition for Hormonal & Sexual Balance** – Eat whole, organic foods; minimize processed sugar; include healthy fats that support hormone production (avocado, coconut oil, olive oil).

2. **Endocrine Detox** – Reduce plastics, switch to fluoride-free toothpaste and filtered water, use clean body products.

3. **Digital Detox** – Limit daily screen time, especially before bed, to reset dopamine sensitivity. Replace with creative, sensory-rich, or nature-based activities.

4. **Grounding & Sunlight** – Spend at least 20 minutes a day in direct natural light and barefoot on earth to restore hormonal and energetic balance.

5. **Movement That Integrates** – Practices like qigong, yoga, or dance blend strength (masculine) with fluidity (feminine) and enhance blood flow to the pelvic area, increasing vitality.

6. **Sacred Stillness & Active Creation** – Alternate meditation (receptive feminine) with focused action (directive masculine) to keep both energies alive.

7. **Pleasure Without Pressure** – Engage in touch, breathwork, or sensual movement without the goal of orgasm, allowing the nervous system to re-learn trust and pleasure without performance anxiety.

When we consciously nurture both energies, reduce overstimulation, and reconnect with the natural rhythms of the earth, we step into a state of internal union—able to love deeply, act decisively, and move through the world as **whole, integrated beings** whose sexual energy is a life-giving force, not a source of confusion or depletion.

Chapter 8:
Rituals, Breath, and Frequency Awakening the God Spark in Your Cells

Some of the practices in this chapter may feel familiar. That is intentional. Certain truths bear repeating—not as redundancy, but as reinforcement—because they are the cornerstones of living in alignment with your highest self. These are not concepts to read once and set aside; they are daily choices that rewire your body, mind, and spirit to operate at their fullest potential.

The body is more than a vessel—it is a living temple, carrying within it the God Spark: a blueprint of divine perfection encoded in every cell. This spark is not dormant by design; it is simply waiting for alignment. When we engage in daily rituals that honor both the physical body and the spiritual essence, we bring our cells into harmony with the original pattern of creation.

This is not about chasing a "higher frequency" for its own sake. It is about aligning with the purest, most coherent frequency your body and soul can hold—so your highest self can work through you with clarity, vitality, and joy. In this state, your mind sharpens, your heart opens, and your actions naturally flow in service to your true purpose.

From the moment you wake, the choices you make either cloud or clarify this inner light. Sunlight on your skin recalibrates your circadian rhythms. Breath practices awaken the nervous system and infuse your bloodstream with life-giving oxygen. Sound—whether through humming, chanting, or sacred frequencies—tunes your cells like a finely crafted instrument. The earth itself offers a natural current of electromagnetic balance, restoring harmony each time you ground your bare feet in its soil.

When these practices are woven into your daily life, they do more than "boost energy" or "relieve stress." They restore the original dialogue between your body and the Divine. They signal to your cells: *It is safe to shine. You are supported. You are home.*

In the pages ahead, we will revisit and refine the essential daily frequencies—physical, energetic, and spiritual—that bring your God Spark fully online. Think of this chapter as your ready-to-use blueprint for living as a coherent, radiant being whose presence uplifts and harmonizes the world.

God Spark Morning Frequency Flow

(A Daily Practice for Cellular Alignment)

This sequence is designed to awaken the **God Spark** in your cellular structure each morning, aligning your body, mind, and spirit with your highest frequency so your work, creativity, and happiness are fully supported.

1. **Lemon Water** – Upon waking, drink a full glass of pure water with fresh lemon. This hydrates, alkalizes, and signals your cells to activate metabolic cleansing.

2. **Essential Oils** – Inhale or anoint the body with a high-frequency essential oil. Choose intuitively—frankincense for spiritual connection, peppermint for mental clarity, citrus for joy. *Affirmation:* ***" I Am the Light That God Shines Through. I release that which no longer serves my body, mind, spirit, health, wealth and happiness."***

3. **Sunlight** – Step outside within the first 30 minutes of waking. Allow sunlight to reach your skin and eyes (without glasses or windows between you and the light) to reset circadian rhythms and stimulate hormonal balance.

4. **Grounding** – Place bare feet on the earth, grass, or sand for at least 3 minutes. This discharges static electrical stress from the body and brings your frequency into coherence with the Earth's Schumann resonance.

5. **Movement** – Engage in gentle stretching, yoga, or slow strength-building movements to open energy channels and oxygenate the blood. And always, walk.

6. **Sound** – End the sequence with humming, chanting, toning, or playing a frequency-aligned sound (e.g., 432 Hz, Tibetan bowls). This stimulates the vagus nerve, harmonizes your nervous system, and imprints the body with coherent vibration for the day ahead.

Bring the Healing Home

Click the QR code below to go directly to Dr. Trish's hand-picked ***Young Living Medical Grade Essential Oils—the same sacred oils I use every day for alignment, clarity, and healing.***

- Uplift your frequency.
- Annoint Your body and spirit.
- Step into your own God Spark with confidence and courage.

Why This Matters:

This flow is not just "healthy habits." Each step directly interacts with your **biophotonic field**, your nervous system, and your energy centers—bringing them into alignment with the **divine blueprint encoded in every cell**. Over time, this creates a baseline state of coherence, resilience, and joy that makes higher states of consciousness your *normal.*

Breath Practices for Energy Activation

Breath is the bridge between body, nerves and spirit—one of the fastest ways to awaken dormant energy, shift emotional states, and align with the God Spark within. These techniques are simple yet profound, and when practiced daily, they amplify vitality, clarity, and spiritual connection.

- **Priming Breath** *(Energy Surge)*
    - Stand or sit tall.
    - Inhale sharply through the nose, lifting the chest and expanding the ribcage.

- Raise arms overhead to the sky and bring down with each breath
- Exhale forcefully through the mouth, drawing the navel toward the spine.
- Repeat 20–30 quick cycles, then pause in stillness, feeling the tingling life force throughout your body, especially your hands.
- *Purpose:* Charges your nervous system, oxygenates blood, and activates mental clarity.

- **Vagus Nerve Breath, Somatic, Holotropic Breath** *(Calm and Coherence)*
  - Inhale gently through the nose for 4–5 seconds, expanding your belly first, then your chest.
  - Exhale slowly for 6–8 seconds, humming softly or making a long "Ahhh" sound.
  - Continue for 3–5 minutes.
  - *Purpose:* Stimulates the vagus nerve, slows heart rate, reduces stress hormones, and restores emotional equilibrium.
- **Wim Hof-Style Breath** *(Deep Activation)*
  - Take 30 deep inhales and relaxed exhales—breathing fully in, letting go without force.
  - After the last exhale, hold your breath until you feel the urge to inhale.
  - Inhale deeply, hold for 10–15 seconds, then release.

- o Repeat for 2–3 rounds.
- *o* *Purpose:* Increases oxygen saturation, strengthens immune response, and can induce a bliss-like meditative state.

Sacred Reminder:

Breath changes chemistry in seconds. You can go from fatigue to focus, from anxiety to deep peace, or from heaviness to lightness simply by changing your breathing pattern. With consistent practice, your breath becomes a tool for instant recalibration of your frequency.

Sound as a Tool to Stimulate the Body's Natural Healing Systems

You've already met the healing power of sound in earlier chapters—through frequency medicine, music as a tuning fork for the soul, and the resonance of ancient instruments. Because this tool is so essential to **aligning with the God Spark in your cellular structure**, we're revisiting it here in a **condensed, practice-ready form**—so you can apply it instantly, without needing to flip back through pages.

Sound is more than something you hear—it's vibration that travels through bone, tissue, and fluid, awakening the body's natural repair systems. Specific frequencies (measured in hertz, Hz) can steady brain waves, soothe the nervous system, and encourage cellular renewal. **Alpha waves** (8–13 Hz) support calm focus, **theta waves** (4–8 Hz) invite deep relaxation and creativity, and **40 Hz** pulses are studied for neuroplasticity and memory. Music tuned to **432 Hz** often feels more heart-centered and grounding; **528 Hz** is linked with cellular harmony and DNA repair.

Your Voice as Medicine: Humming, Chanting, Mantras

When you make sound, vibration originates in the **vocal folds** and travels upward through the **throat** and **nasal passages**, stimulating **tiny hair-like structures (cilia)** that line these areas. These cilia are sensitive to vibration and motion; when activated, they signal the brain to release **beneficial neurotransmitters** such as dopamine, serotonin, and nitric oxide. Nitric oxide production, in particular, increases by as much as 15-fold during humming, supporting circulation, immune defense, and oxygen delivery.

The vibration also resonates in the **sinus cavities**, **chest**, and **heart center**, creating a rhythmic massage for the vagus nerve and influencing heart-rate variability. Over time, this can lead to a calmer nervous system, sharper mental clarity, and a more coherent emotional state.

- **Humming** (closed-mouth "mmm") is the simplest entry point—feel the buzz in your face and chest.
- **Sustained tones** and **chant** (e.g., *Om*, *Amen*, *Alleluia*) deepen breath and unify brain and body rhythms.
- **Mantras** combine sound and intention, anchoring the mind while sending harmonious patterns through the body:
- Christian: *"Amen," "Kyrie eleison,"* or the Jesus Prayer
- Universal/Eastern: *"Om," "So'ham," "Gayatri"*
- Seed sounds for chakras: *"Lam, Vam, Ram, Yam, Ham, Om"*

Ancient Wisdom & Modern Frequency

Throughout history, spiritual traditions have used sound as medicine—Gregorian chant, Vedic mantra, Indigenous drumming, Tibetan bowls—all crafted with precision to guide the body back to balance. Today, modern tools build on that wisdom:

- **Solfeggio tones** (528 Hz for cellular renewal, 639 Hz for connection, 741 Hz for detox, 852 Hz for intuition)
- **Binaural beats** for brainwave entrainment
- **432 Hz** music for natural resonance with Earth's rhythms
- **Tuning forks** and **vibroacoustic sound beds** for direct body stimulation

All of these methods share one aim: restoring coherence so your body's natural healing systems can function without interference.

For expanded detail on each frequency, see **"Using Music to Boost Frequency"** earlier in this book.

God Spark Sound Practice (3–6 minutes)

1. **Hum to Prime (1–2 min):** Sit upright. Inhale gently, then exhale with a steady "mmm." Feel the vibration through your nose, cheeks, throat, and chest. Imagine the cilia waking up and signaling your brain to flood the body with restorative chemistry.

2. **Chant to Cohere (1–2 min):** Choose a single word or mantra (*Om*, *Amen*, *Alleluia*). Inhale fully, then release the sound in a long, resonant tone. Feel it connect your breath, heartbeat, and mind.

3. **Frequency Bath (1–2 min):** Play a healing tone (528 Hz or 432 Hz) from a bowl, tuning fork, or playlist. Sit quietly, breathing 5-in/5-out, letting the sound move through you as if your entire body were "listening."

Tip: This is most powerful when done in the morning sunlight or before rest—allowing the combination of light and sound to align your inner rhythms with the natural world.

Sacred Practice: God Spark Sound Reset (3–6 minutes)

1. Hum to Prime (1–2 min)

- Sit upright. Inhale gently, then exhale with a steady "mmm."
- Feel the vibration through your **nose, cheeks, throat, and chest**.
- Imagine the tiny **cilia** in your nasal passages and throat waking up and triggering a **chemical cascade**—nitric oxide, dopamine, serotonin—flooding your system with repair signals.

2. Chant to Cohere (1–2 min)

- Choose a mantra (*Om*, *Amen*, *Alleluia*).
- Inhale fully, release the sound in a long, resonant tone.
- Feel your breath, heartbeat, and mind syncing into one rhythm.

3. Frequency Bath (1–2 min)

- Play a healing tone (**528 Hz**, **432 Hz**, or Solfeggio sequence).
- Sit quietly, breathing **5 counts in / 5 counts out**, letting the sound move through your cells like light in water.

Best times: Morning sunlight for activation or before sleep for deep restoration.

See "Using Music to Boost Frequency" earlier in this book for deeper exploration of each frequency and how to create your personal playlist.

SECTION 3:
Embodying the Lighthouse: Becoming a Beacon

GOD SPARK

"A Journey into the Light
You Were Born to Embody"

There comes a point in the awakening journey where it is no longer about healing the past—but embodying the light you've already remembered. Section 3 is where that turning point becomes active.

Rather than pushing or striving for transformation, this phase of the journey is about **becoming a living transmission**. You've cleared, remembered, and recalibrated. Now, it's about living from that place—*shining instead of chasing, connecting instead of convincing.*

This section introduces the idea of the human being as a **resonant frequency field**—a bioenergetic lighthouse. It weaves together the wisdom of *The Untethered Soul*, the energetic science of **emotional detachment**, the metaphysical architecture of **Reality Transurfing**, and the deeper StarShine principle:

"Connect, don't convince. Choose love, not fear."

As you will see, **true magnetism is not about trying harder**—it is the natural byproduct of emotional mastery and vibrational coherence.

You become magnetic when you release the need to control outcomes, drop attachments to identity, and let go of the emotional tethers that keep you cycling in lower frequencies. This isn't detachment in the cold sense—it's sacred alignment. You become available to higher love, deeper truth, and clearer guidance.

As I have mentioned before because of its importance; inspired by the **Tibetan Himalayan principle of K'garnae**, this section invites you to **declare your intention** from a place of soul authority, but remain unattached to how or when it manifests. You are not defined by your goals, or even by their attainment. You are defined by your alignment.

You'll also explore how the **dopamine-fueled loops of modern society**—addiction to outcome, overstimulation, emotional reactivity—drain your magnetic field. By shifting from external striving to internal stability, you begin to **attract through resonance**, not through effort.

This section will show you how to:

- Clear emotional attachments that leak energy
- Reclaim sovereignty over your internal state
- Live as a magnetic force without force
- Let your **presence become your message**
- Use **intention without obsession**, clarity without control

As you deepen into this embodiment, you move from *student to signal*, from *seeker to seer.*

You become the lighthouse.

Chapter 9: Emotional Mastery and the Untethered Soul: Releasing Attachments

As we begin this next section, we enter the territory of **emotional mastery**—a sacred phase of becoming that turns inward, where we learn not just to heal or rise, but to remain clear and free even when life pulls at us with intensity. If you've made it this far, you already know that everything in your external reality reflects an internal state. This chapter gives you the keys to staying steady when waves of emotion or outside circumstances challenge your peace.

Many of the ideas here have been touched on throughout earlier chapters—because **this is the work that transforms every other area of life**. Yet, now we go deeper, more directly, into the mechanics and mysteries of attachment, identity, and the untethering that leads to personal freedom.

The inspiration for this chapter draws heavily from *The Untethered Soul* by Michael A. Singer, whose simple yet profound guidance reminds us that **you are not your thoughts or emotions—you are the one who observes them**. This awareness is the gateway to freedom. When you shift your identity from the one who is hurt, worried, or reactive, to the one who simply sees, feels, and breathes through it all—you begin to reclaim your power and your peace.

In spiritual terms, this is the moment you stop riding every wave and start becoming the ocean. In energetic terms, this is where your frequency stops collapsing under external conditions and begins generating its own steady light. **It's the training of the witness**, the capacity to remain conscious and open in the presence of discomfort—without collapsing, numbing, or pushing it away.

This practice is not about detachment in the cold sense, but about **expanding your ability to stay present** without needing to control, grasp, or resist. When we talk about "emotional mastery," we are not suppressing the emotional body—we are honoring it, learning to read it, and choosing how to respond instead of reacting.

Just as a lighthouse doesn't chase ships, the awakened soul doesn't chase approval, safety, or even healing. Instead, it becomes the source of clarity that others naturally orient toward.

How Attachment Drains Frequency

Attachment isn't love. It masquerades as care, loyalty, or investment—but often, it is **entanglement with a fixed outcome, unhealthy addiction, fear, or identity**. From an energetic perspective, attachment pulls your frequency downward because it anchors your nervous system to something outside your control: a person, a belief, a timeline, a story.

When you're attached, your energy is no longer circulating freely within you—it is **looping into an external anchor**, creating subtle or intense dependency. This can be especially deceptive when the attachment appears spiritual or noble: a desire to serve, to save, to be seen, or even to heal someone else.

But every time you grasp too tightly, your frequency constricts.

The Biology of Attachment

From a physiological standpoint, attachment creates **dopaminergic loops** in the brain. The anticipation of a particular outcome—like being chosen, being right, or feeling safe—releases neurotransmitters that reward the emotional investment. But when the outcome doesn't arrive, cortisol and adrenaline spike. This keeps your body on high alert, feeding cycles of stress, craving, and depletion.

In contrast, true presence releases oxytocin, serotonin, and endorphins—chemicals of trust, joy, and contentment. These emerge **when you're rooted in the moment**, not grasping for the next.

The Spiritual Cost

Spiritually, attachment blocks the flow of divine will. It says, "I know how it's supposed to go," which resists the mystery. It confuses the soul's longing for connection with the ego's desire for control. And over time, it dims the very light we're trying to share.

As one of our **StarShine Guiding Principles** affirms: *Connect, don't convince. Choose love, not fear.*

Attachment is a fear-based grasping—often stemming from the illusion that we are separate from love, from Source, from our true path. It binds your frequency to the lower vibration of anxiety, control, and conditional self-worth.

When you notice yourself tightening, chasing, proving, or waiting for something outside yourself to validate your worth—pause. Breathe. Let go. Not from apathy, but from wisdom.

K'garnae, the ancient Tibetan-Himalayan principle, teaches: *Declare your intention, then release it.* Do not define yourself by the goal or the outcome. This is not passivity—it's power. It is living in alignment with the truth that your essence is enough, even while you grow, serve, or m**anifest.**

Practical Tools for Emotional Detachment Without Disconnecting from Love

Letting go does not mean withdrawing love. In fact, detachment—when practiced consciously—is the ultimate act of love. It frees both you and others to be seen clearly, accepted fully, and loved unconditionally without grasping, controlling, or projecting.

Emotional detachment is not about shutting down or becoming indifferent; it's about creating enough internal spaciousness to observe your reactions without being ruled by them. It is standing in your truth without needing to convince, control, or cling. It's the clarity of connection without cords.

Here are several **sacred practices and frequency-based tools** to help you cultivate emotional detachment while staying rooted in love:

1. Return to Presence Through the Body

The body is the anchor point for the present moment. When emotion floods the system, place your hand on your heart and say aloud: *"I am safe. I am here. I am whole."*

Feel your feet on the floor, your breath in your chest, the space around your body. This brings your energy back into alignment with what is real—not imagined fear or future longing.

2. Cut Cords with Compassion

Visualize any energetic attachments like cords between your solar plexus and the other person or situation. With love, visualize gently unhooking these cords and returning the energy to its rightful source—blessing it with peace. This doesn't sever the relationship; it restores energetic sovereignty. Use Young Living Essential Oil Lime on your crown of your head.

3. Practice Non-Attached Intention (Inspired by the Tibetan K'garnae Principle)

Declare your intentions with clarity and full commitment—but hold them lightly. Trust that your desire is already known to the universe. Let go of defining yourself by whether or not it happens. You are not your goals. You are the soul behind them.

4. Reframe the Narrative

Catch yourself when you start looping stories: *"They always do this,"* or *"If I had just..."* or *"I'll never have..."*

Replace these with *"I am learning to trust the process,"* or *"Everything is unfolding in alignment with my highest frequency."* Thought is frequency—choose ones that elevate.

5. Bless, Don't Beg

When someone or something leaves your life, bless it. Don't beg it to return. Begging operates from fear; blessing activates flow. Say aloud, *"I release this with love and open my hands for what's next."*

6. Practice Daily Energy Hygiene

Just as you shower daily, clear your emotional field daily. Use essential oils, grounding in nature, sound frequency (like 528 Hz or 741 Hz), or breathwork to cleanse emotional residue that isn't yours. This keeps your aura and nervous system from storing and reactivating past emotional loops.

7. Connect Without Convincing

STARSHINE's Guiding Principles: *Connect—don't convince.*

When you're tempted to prove your worth or perspective, pause. Love doesn't require defense. When you stay in your own light, those meant for you will see you without explanation.

8. Anchor to the God Spark Within

Your worth, power, and direction come from the Divine spark inside you—not from approval, attention, or outcomes. Return to this flame daily through prayer, stillness, or joyful creation. Let it remind you that nothing outside you can define or diminish your frequency.

Detachment is not the absence of feeling; it is the presence of trust. When you release the grip on what you cannot control, your heart opens more fully, and your frequency rises into radiant coherence. From here, you don't just love—you *become* love.

Chapter 10: The Magnetic Effect of High Frequency

By now, we've explored how emotions, thoughts, trauma, breath, and even sound shape our internal frequency—and how that frequency affects our experience of the world. In this next phase, we shift from inner calibration to outer influence.

This chapter is not about doing more. It's about becoming more—more coherent, more aligned, more true to the divine signal within you. Here, we explore the radiant field you emit when you live from your God Spark—and how this high-frequency embodiment becomes magnetically attractive to the experiences, people, and opportunities meant for you.

This is the heart of spiritual magnetism.

True manifestation doesn't happen through frantic effort or obsessive control. It occurs through **resonance**—when your energetic field harmonizes with what you're calling in. It is less like chasing and more like *sounding a note so clear that everything in tune with it begins to arrive.*

High frequency doesn't just feel good—it reorganizes your life.

In this chapter, we'll explore three key principles:

- How resonance creates faster and more authentic manifestation than striving.
- Why alignment naturally draws people and opportunities toward you.
- And how to live as a lighthouse—emitting light rather than chasing after ships.

This is where effort gives way to radiance.

Let's begin.

Resonance vs. Effort in Manifestation

Manifestation, in its highest form, is not about control—it's about coherence. When you force outcomes through willpower alone, you're often operating from lack, fear, or disconnection. The effort is coming from a frequency of *not having*—and that very signal keeps what you desire just out of reach.

Resonance, on the other hand, works on a deeper, more elegant level. When your inner frequency matches the frequency of what you're calling in, manifestation becomes almost effortless. You don't *pull* it in—it *arrives* because it belongs.

In physics, resonance occurs when one object vibrating at a certain frequency causes another object to vibrate at that same frequency. Tuning forks demonstrate this beautifully: when one is struck, others nearby begin to hum in harmony—not by effort, but by coherence. The same is true for humans. When your field is clear and aligned, you become a tuning fork for divine design.

This is why clarity, joy, gratitude, integrity, and spiritual alignment are so powerful—they are high-frequency states that shape reality faster and more sustainably than hustle or grind.

Rather than pushing, think of yourself as attuning.

Attune to the vibration of love, and you attract people who are open-hearted.

Attune to abundance, and pathways begin to open that logic alone could not orchestrate.

Attune to truth, and synchronicities emerge to guide your next steps.

Manifestation is not magic—it is alignment. When you remember this, you can stop exhausting yourself trying to prove, earn, or chase. You become the attractor by becoming the match.

Why Alignment Attracts Faster Than Striving

Striving is fueled by force—it pushes, grasps, and often works against the natural current. Alignment, by contrast, is receptive. When you are aligned with your deepest truth, you no longer chase outcomes; you become the space through which outcomes arrive. This shift accelerates results because the energy once lost in resistance is now freed to magnetize exactly what matches your frequency. In this way, alignment is not about doing more—it is about allowing more.

The Physics of Resonance and Divine Order

In the natural world, **alignment isn't passive—it's magnetic.** From the orbit of the planets to the bonding of atoms, **the universe is held together not by forceful striving, but by invisible laws of resonance, charge, and attraction.** The same is true in your own body—and in your ability to manifest a harmonious life.

The Body as an Electromagnetic System

Your body generates **both electric and magnetic fields**—especially from your heart and brain. The **heart's electromagnetic field is 5,000 times stronger than the brain's** and can extend several feet beyond the body. When you're aligned—when thoughts, emotions, and actions are in coherence— this field becomes stronger, more coherent, and more attractive. It literally broadcasts a signal into the quantum field.

Living as a Lighthouse Instead of Chasing Ships

Repeating this important concept can't be read too many times: A lighthouse does not wander the seas in search of vessels; it stands rooted, radiant, and unshakable in its purpose. In the same way, your task is not to chase opportunities, people, or outcomes—it is to embody such coherence of light that what is meant for you navigates toward you with certainty. When your inner field is strong, you no longer need to pursue; you become a fixed point of guidance, safety, and recognition. Ships find the shore not because the lighthouse moves, but because it shines. Likewise, your life aligns when you allow your energy to broadcast steadily, without desperation or pursuit.

The Power of Becoming Rather Than Pursuing

In a world driven by hustle, many are taught to chase—people, goals, validation, even love. But **lighthouses don't run up and down the shore trying to catch boats.** They **stand still, rooted, and radiate light.** It is their unwavering presence—not their effort—that **draws ships safely to shore.**

This is the essence of high-frequency embodiment. When you become the vibration of what you desire, you don't need to convince or chase. You magnetize. You become a field of coherence.

As the poet Rumi wrote:

"What you seek is seeking you."

Like a lighthouse, your embodied alignment emits a steady frequency—clear, unwavering, and true. This frequency calls in what is already resonant with your soul's path. It is not passive, but profoundly magnetic. No chasing, convincing, or striving is necessary—only the courage to stay lit, stay grounded, and stay true.

SECTION 4:
Igniting the World: Spreading Light, Love, and Truth Through Action

You are not here to stay hidden.

You didn't come to Earth to keep your light tucked safely inside the walls of meditation rooms, journals, private breakthroughs, private jets, or luxury yachts.

You came to **radiate**—to let the **God Spark** within you become a living frequency of love, truth, and transformation in the world.

This final section is about **integration and ignition**.

It's about turning everything you've remembered—every alignment, every healing, every activation—into **embodied action**. Not striving. Not pushing. But **living from your frequency**, so purely and powerfully that others remember their light in your presence.

It begins with **relationships**. Because if we cannot live what we know in our most intimate connections, we will never change the world. The way we love, listen, partner, speak, and stand beside one another **is the message.**

This is where healing becomes service.
This is where frequency becomes legacy.
This is where your God Spark becomes a lighthouse for others.

Chapter 11: The God Spark in Relationships and Sacred Love

Holy Love, Divine Union

There is a point on the journey when your light no longer simply returns to Source—it begins to seek another flame to dance with. This is not about dependency or desire, but resonance. Sacred love marks the moment when two sovereign beings recognize the divine spark in one another and consciously choose to amplify it—together.

As your frequency rises and the veils of illusion dissolve, your very biology begins to shift. Your cells emit more light. Your eyes become literal windows of the soul—projecting not just emotion, but **photonic pulses** that science now confirms radiate from the body. In moments of divine connection, Zero Point bursts occur—measurable surges of coherence, emotion, and electromagnetic harmony that strengthen both the heart field and the body's ability to self-heal. This is not metaphor. It is mystery made visible.

When sacred union is real, you can see it. You can *feel* it. The eyes glisten with knowing. The aura expands. The body softens. And healing accelerates, as love itself becomes the medicine.

This chapter is the gateway to co-illumination—a remembrance that **you were never meant to hold your light alone**. Just as the Earth contains invisible grid lines of divine current, so do we carry unique soul signatures designed to align with others. Sacred relationship, when entered with reverence, becomes a **conduit for greater light**—not just for the individuals involved, but for the world.

Here, we explore how divine union transcends attachment. How shared spiritual practice creates a third field of energy. And how unconditional support, open-hearted communication, and deep soul recognition activate not only personal evolution—but planetary healing.

This is sacred love: not a sanctuary to hide in, but a frequency to *broadcast from*. The God Spark ignited between two aligned souls becomes a beacon—an anchor for truth, grace, and transformation.

Let us step into this divine mirror now. Let us remember who we become in the presence of a love that reflects the light we almost forgot we carried.

The Chemistry of Sacred Love

While we may speak of love in celestial tones, the experience of it is grounded in the exquisite orchestration of our biology. Sacred love engages not only the spirit—but also the bloodstream.

- **Oxytocin**, the bonding hormone, flows during moments of physical touch, deep intimacy, and even eye contact. It softens the heart and reinforces trust.

- **Dopamine** provides the fuel of passion and anticipation. It makes you light up with joy at a message or smile from your beloved.
- **Serotonin** fluctuates, especially in the early phases of sacred connection, often bringing a laser-like focus on the beloved's essence.
- **Endorphins and Vasopressin** increase feelings of safety, pleasure, and deep attachment, especially as relationships mature.

In sacred love, these natural chemicals don't just bind us—they *tune* us. They initiate a kind of biochemical alchemy that aligns us to something higher than instinct: soul resonance.

The Frequency of Sacred Love

Beyond the biochemical, sacred love reveals itself as vibration—subtle, radiant, and deeply felt. Love is a frequency that can be raised, refined, and radiated.

- The **Heart Chakra** is the energetic center of divine connection. When open, it sends and receives frequencies of forgiveness, compassion, and oneness.
- **639 Hz** is often associated with relationship harmony and healing, said to balance interpersonal dynamics and clear emotional residue.
- **528 Hz**, the so-called "miracle frequency," is linked with DNA repair, emotional restoration, and spiritual awakening.

While science continues to catch up to what the mystics have long known, personal experience often validates these vibrations. Sacred love feels expansive. It carries warmth, synchronicity, and a sense of timeless truth. It elevates everything it touches.

The Fusion Point: Chemistry Meets Frequency

The most profound relationships are not one or the other—they are both. The sacred meets the cellular. The invisible meets the hormonal.

As you open your heart and body in alignment with sacred intention, the body responds. Zero Point bursts—those pure, potent pulses of possibility—may be felt during connection, eye-gazing, union, or even mutual prayer. In these moments, new neural pathways form. A kind of healing begins that transcends talk or logic.

This is why **sacred love can feel like coming home**—and why its impact can linger for lifetimes. When frequency and chemistry align, we are transformed at every level. This is why synchronicity starts to be the normal of experiences.

Cultivating Sacred Love as Spiritual Practice

Sacred love does not arise by accident. It is summoned, sculpted, and sustained. Here are foundational elements:

- **Self-Awareness and Acceptance**: Sacred love begins within. The more whole you become, the more whole the connection becomes.
- **Openness and Vulnerability**: The true beauty of love is revealed in rawness. Sacred connection requires us to be seen.
- **Unconditional Presence and Support**: Not just being there—but truly *holding space* for the other's soul to unfold.
- **Mindful Communication and Reverent Touch**: Every word, every gaze, every embrace becomes a sacred act.

- **Honoring the Third Energy**: In sacred union, a third presence is born—the relationship itself. Treat it as a living, breathing soul.

When approached with reverence and grace, sacred love becomes a **living technology**—a healing force, a catalyst for awakening, and a container for miracles.

The Difference Between Attachment and Divine Union

As we move deeper into the realms of sacred love, one of the most pivotal awakenings is the distinction between **attachment** and **divine union**. Though they may outwardly appear similar—both filled with passion, intensity, and connection—their roots, fruits, and frequencies are entirely different.

Understanding Attachment: The Illusion of Possession

Attachment, at its core, is a connection forged through the lens of the ego. It stems from the belief that *something outside of us* can complete us, fulfill us, or make us whole. This longing, though often masked as love, is usually driven by fear, insecurity, and a desperate search for control.

1. Ego Driven and Fear Based

The ego thrives on identity, and attachments often form when we define ourselves by our relationships, roles, or possessions. This leads to fear of loss—of status, love, comfort, or identity. We begin to clutch tightly, mistaking the temporary for the eternal.

"If I lose them, I'll lose myself," becomes the unspoken mantra of attachment.

This fear creates a subtle or overt desire to *possess*, to hold on, to make permanent something that is inherently fluid.

2. Possessiveness and Control

Attachment often manifests as jealousy, possessiveness, or the need to manage how others behave, love, or show up. There is a transactional quality: *I give, so that you stay.* It becomes less about free will and more about dependency.

In truth, attachment seeks not union—but **ownership**.

3. Conditional and Transactional

Attachment is conditional: *I love you because you make me feel [desired, safe, whole].* This isn't inherently wrong—it's simply human. But when left unchecked, it becomes limiting and burdens the relationship with expectations, demands, and unconscious agreements.

When that person or thing no longer fulfills our inner need, we may withdraw love, become resentful, or spiral into emotional disarray.

4. Hindrance to Spiritual Growth

SpiritualDirection.com and Catholic teachings echo the warning that attachment can tether us to the material plane. St. John of the Cross describes the soul's journey as an **ascent away from earthly attachments** toward union with God. The more we grasp, the more we are held back.

As long as our hearts are entangled in things that perish—outcomes, recognition, specific people—we are blocked from entering the expansive flow of divine love. Attachment is like tying your spirit to a boulder and then asking it to fly.

5. Examples of Attachment

- Clinging to a romantic partner who no longer aligns with your soul's path, simply because of shared history or fear of being alone.
- Obsessing over material wealth, seeing it as your only source of power.
- Becoming addicted to validation or success, and crumbling when they are absent.
- Needing your partner to behave or change in order for you to feel loved or secure.

These examples aren't moral failings—they're part of the human experience. But they become obstacles when mistaken for love.

Embracing Divine Union: The Sacred Mirror

Divine union, on the other hand, is not rooted in need—but in **remembrance**. It is a recognition of shared divinity, a sacred echo that calls forth the best in each soul. In divine union, the other is not your source—but your mirror.

1. Selfless and Soul-Centered

Divine union is built on **selfless love**—not the kind that erases you, but the kind that expands you. It is not about *getting*, but about *being*. It emerges when both individuals are deeply connected to their own essence and allow their connection to magnify their sacred paths.

The soul doesn't cling—it communes.

2. Expansive and Elevated

Instead of shrinking to fit another's needs or demands, divine union expands both partners. You feel more you in their presence—not less. There is no need to hide, shrink, conform, or perform. The relationship becomes a sacred container where transformation unfolds in mutual freedom.

This kind of love elevates consciousness and opens channels of creativity, healing, and purpose.

3. Unconditional Love and Acceptance

There is no *"I'll love you if…"*—there is only *"I see you. I honor you. I walk with you, even if our paths change."* Divine union is steeped in the frequency of **unconditional love**—the kind that reflects God's nature. It does not control, it does not punish, and it does not demand proof.

This is why divine partnerships often feel like home: they offer **soul safety**.

4. Liberation Through Love

Far from making us feel caged, divine union sets us free. The bond is not a shackle—it is a **sacred flame** that illuminates each being's individuality while guiding them toward a shared divine mission.

This freedom creates space for joy, sexual and emotional vitality, spiritual evolution, and authentic expression.

5. Examples of Divine Union

- A spiritual partnership where both individuals support each other's path without needing to "complete" one another.

- A deep soul connection where presence, not performance, is enough.
- A love that endures even when physical presence doesn't—because it's rooted in energetic truth, not circumstance.
- A communion with God or the Divine that brings clarity, peace, and profound belonging.

These connections transcend time and space. They are portals of remembrance.

Key Differences at a Glance

Aspect	Attachment	Divine Union
Rooted In	Ego, fear, lack	Love, freedom, wholeness
Goal	Possession, validation	Expansion, communion
Frequency	Contractive, anxious	Expansive, peaceful
Love Type	Conditional	Unconditional
Energy Dynamic	Controlling, needy	Free, supportive, honoring
Impact	Draining, stagnating	Nourishing, catalyzing
Temporal Nature	Finite, changing	Infinite, enduring
Path Outcome	Stagnation, dependence	Liberation, transformation

Shedding Attachment, Stepping into Union

The path to divine union begins with the courageous work of **detachment**—not to disconnect from life or love, but to release illusions.

As the mystics say: *Detach not from love, but from illusion.*

This is echoed across traditions—from Buddhism's call to release craving, to Sufism's union with the Beloved, to Christ's teaching that the Kingdom is within. Each points to the same truth: divine love is already here. We need only clear the clutter of dependency and projection to access it.

Practices for Transcending Attachment:

- **Self-Inquiry**: Ask yourself, *Do I love this person (or thing) for who they are, or for how they make me feel?*
- **Emotional Awareness**: Notice when fear or control creeps in. These are indicators of ego attachment.
- **Spiritual Alignment**: Ground in your own connection to the Divine first. No one else can be your source.
- **Sacred Solitude**: Learn to be alone without being lonely. Divine union begins within.
- **Affirmation Practice**: "I release all illusions of control. I open to love that liberates, not binds."

From Earthly Bond to Cosmic Dance

Facebook's reflections on divine union beautifully frame it as something that **transcends possessiveness and self-interest**, becoming a co-creative dance that honors each soul's divine timeline.

True union, it teaches, is not about *holding on*—but about *holding sacred.*

Muslim Space emphasizes a similar principle: that attachment, while natural, must be purified in order to reach higher stations of love. Love of the Creator—through and within creation—is the ultimate aim. Every human love becomes a **mirror of the One**.

A Final Reflection

Attachment may be part of our human learning—but divine union is part of our divine memory. One binds. The other frees. One contracts. The other expands. One feeds the ego. The other feeds the soul.

And sacred love? Sacred love chooses divine union over attachment, again and again. It recognizes the beloved not as possession, but as portal. It is not about what you can get—but what you can become **together**.

As you walk this path, may your relationships evolve from grasping to grace—from craving to co-illumination—from fleeting infatuation to **eternal communion**.

Let every connection be a temple.
Let every love be a prayer.
Let every soul you meet remind you:
You are already whole.

"Love is Not Possession" (Inspired by Kahil Gibran)

Love one another, but make not a bond of love:Let it rather be a moving sea between the shores of your souls.
Fill each other's cup but drink not from one cup.
Give your hearts, but not into each other's keeping.
For only the hand of Life can contain your hearts.

This poetic passage from **Kahlil Gibran's "The Prophet"** is a sacred anthem for divine union—two souls joined in freedom, not in captivity.

Scripture: 1 Corinthians 13:4–7 (NIV)

Love is patient, love is kind.
It does not envy, it does not boast, it is not proud.
It does not dishonor others, it is not self-seeking,
it is not easily angered, it keeps no record of wrongs.
Love does not delight in evil but rejoices with the truth.
It always protects, always trusts, always hopes, always perseveres.

This is often quoted in weddings, but when read from the lens of divine union, it becomes a vibrational compass of what **God-centered love truly looks like.**

Sacred Song: "Set Me as a Seal" – Song of Songs 8:6–7

Set me as a seal upon your heart, as a seal upon your arm;
For love is strong as death, passion fierce as the grave.
Its flashes are flashes of fire, a raging flame.
Many waters cannot quench love, neither can floods drown it.

This passage—ancient, wild, holy—describes the burning essence of **true divine union**: not possessive, but enduring and sacred beyond time.

Soul Partnerships as Accelerators of Divine Purpose

There is a divine phenomenon that occurs when two souls—vibrating in resonance—enter into sacred union: their **frequency fields merge and amplify**, creating a third energetic vortex far more powerful than either could produce alone. This isn't poetic metaphor; it's measurable energy. It's **Zero Point Energy activated through love**.

In the field of quantum physics, when coherence is achieved between two separate waveforms, **a harmonic resonance occurs**, creating *constructive interference.* In simple terms: **the light gets brighter.** The flow becomes smoother. The impossible begins to feel inevitable.

This is what happens in sacred soul partnerships—not just romantic, but creative, spiritual, or even catalytic connections. When the purpose of the union is **remembrance and contribution**, the universe itself seems to open doors, soften resistance, and conspire in favor of the mission.

Divine Pairings Across Time

Throughout history, we've seen radiant examples of partnerships that ignited not only each other's gifts—but entire **movements of transformation.**

- **Harriet Tubman and William Still**: A divine collaboration that helped hundreds escape slavery, driven by inner calling and trust in divine guidance.
- **Frida Kahlo and Diego Rivera**: Though turbulent, their artistic symbiosis pushed each other into greatness, blending pain and passion into cultural immortality.
- **Marie and Pierre Curie**: Merged intellects and souls who discovered radiation and ushered in a new era of scientific understanding.

These unions—whether romantic, professional, or spiritual—shared a **sympatico** energy field. A kind of sacred rhythm. Each individual brought their own frequency, but the partnership created **a harmonic expansion**—a new field of transformation.

Divine Sympatico: The Zero Point Spark

When soul partnerships align with divine purpose, a **Zero Point coherence** is established. That is:

- Emotional tension dissolves into shared mission.
- Comparison dissolves into co-creation.
- The "I" surrenders to "We"—not from loss, but from *multiplication.*

Zero Point Energy, in its truest essence, is **pure potential—undirected, silent, still.** When two awakened beings join with reverence, purpose, and resonance, they create a **magnetic field** that **pulls from this infinite source**—manifesting, healing, and serving with ease.

This is not ordinary relationship. It's **Divine Geometry in Motion**.

Hallmarks of Soul-Aligned Partnerships

Here are a few signs a relationship is **accelerating your divine purpose**:

- **Synchronicities multiply.** Doors open. Time bends. The right people appear.
- **Shared silence feels rich.** There's no need to prove, only to *be.*
- **Energy expands, not depletes.** You feel **more you**, not less.
- **You become more generous.** Your desire to serve grows.
- **Creative output increases.** Whether it's art, teaching, healing, parenting, or simply living—your light **wants to be given**.

Sacred Reminders

A true soul partnership is **never about rescuing or completing the other**—it's about **revealing each other's light**. Your purpose becomes clearer. Your burdens feel lighter. You rise—together.

And this is not limited to one kind of relationship. You might find your sacred accelerator in:

- A best friend who sees your essence.
- A spiritual teacher or student.
- A beloved pet who holds you in unconditional presence.
- A divine romance that amplifies your entire path.

Sometimes, the most sacred accelerators don't stay forever. They show up, ignite the God Spark within you, and lovingly release you to your next level. That, too, is divine.

Closing Blessing

If you have not yet encountered your soul partner, *trust that they are being magnetized to your frequency now.* Every time you remember who you are, clear the illusions, and radiate your light—you become more visible in the field of divine resonance.

You don't need to chase sacred love.
You need only **become a match to your mission.**
The rest… will arrive in divine timing.

Chapter 12: Purpose, Prosperity, and High-Frequency Wealth

Manifesting from Divine Alignment, Not Driving

There is a moment on the sacred path when the energy of striving gives way to the frequency of receiving. In this shift, the old rules of force, hustle, and control dissolve, revealing a more powerful law—**the law of harmonic resonance.** Just as sound waves align to create symphonies and radio frequencies must be tuned to hear music clearly, so too must our inner state come into vibrational alignment with divine purpose to receive the flow of true wealth.

The Earth is abundant by design. Trees do not strive to grow. They rise, stretch, and unfold with the sun. Mushrooms don't hustle; they emerge overnight through networks of quiet cooperation and intelligent response to the environment. Rivers do not push; they flow with the path of least resistance, yet they carve canyons over time. And so too are we invited—*not to push harder*, but to become more attuned.

This chapter invites you to reclaim wealth—not as a static sum of money or status, but as a **living frequency** that responds to coherence. It asks you to reimagine prosperity through the lens of Tesla, who lit lamps from the air, not by wires or brute electricity, but by **tuning into nature's own invisible, intelligent design.**

Tesla famously said, "If you want to find the secrets of the universe, think in terms of energy, frequency, and vibration." And that is what we'll do here. Because **true wealth** is not amassed; it is attracted. It is not hunted; it is harmonized. It is not bought; it is embodied.

The Invisible Laws of True Wealth

Across cultures and sacred teachings—from the Bible to ancient Vedic texts—we are told that we are heirs to a Kingdom, a divine inheritance. But what does that mean in a world obsessed with scarcity and competition? And how do we bridge the mystical truth of abundance with the lived reality of bills, jobs, and modern pressures?

The answer lies in **shifting the source of your prosperity from external effort to internal alignment**. Wealth becomes a byproduct of integrity, joy, service, and resonance. As you elevate your frequency through gratitude, clarity, and daily sacred practices, you become a *conduit* for divine provision.

When you are aligned with your highest truth, life provides for you. It's not magic. It's magnetic law.

We'll explore how **push energy blocks abundance**, and why relaxing into **joyful contribution** unlocks more prosperity than endless productivity. We'll decode the spiritual technologies behind real wealth, from energetic coherence to Tesla's field resonance, from sacred generosity to the geometric patterns of financial flow.

Tesla, Ether, and the Hidden Field of Supply

Nikola Tesla's inventions were never just mechanical. They were metaphysical. He didn't merely build machines—he unveiled the subtle laws of the cosmos.

Tesla discovered that **energy exists everywhere**, not just in power plants, but in the atmosphere. In one famous experiment, he powered a lightbulb wirelessly—drawing energy from the **etheric field** around him, proving that when the **right resonance and alignment** occur, **light manifests from seemingly nothing.**

That is the energetic mirror of financial prosperity.

You don't need to hustle harder. You need to tune in deeper. Prosperity is already present—an ambient potential surrounding us like the ether Tesla tapped into. When your **thoughts, beliefs, and actions** match the frequency of service, beauty, and trust, **wealth begins to appear**.

Just as Tesla created earthquakes with a tiny device pulsing at a specific vibration—small yet powerful—your *intentional alignment* can shake the foundations of old scarcity thinking and initiate seismic shifts in your life.

The Living Metaphors of Nature

Let's turn to the forest.

In the **Redwood groves of Northern California**, some of the tallest trees on Earth stand not because of individual strength, but because of their **interconnected root systems**. Their roots intertwine, share nutrients, and stabilize one another. Their **health creates resilience**, and their **resonance creates coherence.**

Compare this to a diseased forest. When the soil is depleted, beetles attack weakened trees. The resonance is off. The system collapses.

Prosperity works the same way. In a coherent field of thought, care, giving, and grounded joy, life flourishes. In dissonant energy—fear, hoarding, self-doubt—systems decay.

Rain nourishes the Earth not by force, but by surrender. It falls, and the **electrical ions** in the atmosphere shift. Negative ions increase. The air feels fresh. Plants revive. Rain is not just water—it is atmospheric transformation.

Fire, too, reveals this mystery. In unhealthy forests, fire spreads fast, chaotic, and deadly—because imbalance has created vulnerability. But in healthy ecosystems, fire can be **regenerative**—clearing underbrush, restoring soil chemistry, awakening ancient seeds.

Nature Never Pushes. It Pulses. It Breathes. It Aligns.

So must we.

The Role of Gratitude, Joy, and Coherence

Gratitude, like Tesla's field resonance, is not simply a good feeling—it is a magnetic signal. It tells the universe, *"I'm already rich."* And what's rich expands.

Joy is not frivolous, it is essential. Studies in neuroscience now show that joy expands the brain's creative capacity, problem-solving ability, and emotional resilience. From an energetic standpoint, joy aligns you with the flow of abundance—it clears static from your field and makes you more receptive.

Coherence—when your brain waves, heart rhythms, and actions align—creates what scientists call **constructive interference.** You're no longer working *against yourself.* You become a broadcast tower of peace, power, and receptivity.

In sacred prosperity, money doesn't just come. Meaning comes. Clarity comes. Ease comes.

You become like the waterfall: changing structure as it falls, but always moving in harmony with gravity and grace. You become like the light-bulb in Tesla's hand—suddenly illuminated because the environment was tuned to receive.

Adamantine Particles and the Cosmic Currency of Light

Some spiritual scientists speak of **adamantine particles**—indivisible units of divine intelligence that respond to thought, intention, and love. These particles, they say, are the **substance of creation itself**, flowing through the Zero Point Field, waiting for instruction.

Like seeds in the wind, they gather and become form when aligned with coherent intention.

Tesla tapped the same principle. So did Einstein, in his explorations of **unified fields.** So do you—when you **speak from truth, give with love, and act from alignment.**

These particles are drawn to light. They respond to frequency. They *become wealth, relationships, healing, ideas, opportunities.*

You don't have to chase money. You have to **broadcast truth**.

You don't have to demand supply. You have to **become the channel through which Divine Supply flows.**

And when you stop pushing…

You start receiving.

Why "Push" Energy Blocks Abundance

Let's now explore the subtle yet destructive impact of push energy on your field—and how releasing it activates the flow of Divine Provision.

Some interpretations suggest that an excessive focus on "push energy" or striving for abundance, particularly material abundance, can block the experience of genuine fulfillment and lead to feelings of dissatisfaction and attachment.

Scientific and psychological perspectives

- **Scarcity Mindset:** When we operate from a scarcity mindset, fixating on what we lack, our brains become wired to perceive limitations and focus on lack rather than possibilities. This can hinder our ability to recognize and attract opportunities that could lead to abundance.
- **Stress and Resistance:** Excessive striving and attachment to specific outcomes can create stress, anxiety, and a feeling of resistance when things don't go as planned. This negativity can block a natural flow of opportunities and make us less open to the good things that may be trying to enter our lives.
- **Cognitive Bias:** The brain's natural tendency to prioritize threat and discomfort can make it challenging to shift from a focus on what's missing to a focus on abundance. Research shows that a scarcity mindset can even lower cognitive performance.

Jesus' Teachings

- **Focus on the Kingdom of God:** Jesus emphasizes seeking first the Kingdom of God and his righteousness, with the promise that other things, including necessities, will be provided. This suggests prioritizing spiritual and relational well-being over solely pursuing material wealth.

- **Beware of Greed:** Jesus warns against covetousness and placing too much value on material possessions, highlighting that true life is not measured by what we own. According to many, this warning challenges the idea that wealth is a sign of divine favor and encourages focusing on spiritual richness instead.
- **Abundant Life:** Jesus talks about coming so that we may have life abundantly. This is interpreted as a life filled with joy, peace, love, purpose, and fulfillment, regardless of material possessions or outward circumstances. I believe what some interpretations say, that the Greek word for "abundant" in this context suggests not just quantity but also quality of life.
- **Trust in God's Provision:** Rather than relying solely on our own "push energy" and efforts, Jesus teaches about trusting in God's provision and generosity, believing that He will provide what we need.

In essence, both scientific perspectives on the mind and Jesus' teachings suggest that an overemphasis on striving and accumulating can lead to a narrow, fear-driven focus that blocks the experience of true abundance, which encompasses well-being, contentment, and fulfilling relationships, rather than simply material wealth.

The Role of Joy, Service, and Gratitude in Wealth Creation: Sacred Prosperity Principles

Many wisdom traditions and modern thought leaders suggest that cultivating joy, service, and gratitude are not just feel-good practices, but also powerful principles that can attract and sustain true prosperity, going beyond mere financial accumulation.

Here's how these elements are believed to intertwine:

1. Joy and wealth creation

- **Positive Mindset and Decision-Making:** A joyful and optimistic mindset fosters clarity and expands one's financial time horizon, leading to better planning and investment choices. Research suggests that positive emotions broaden cognitive awareness and curiosity, allowing for a more expansive perspective on life and decision-making, according to Forbes Magazine.

- **Reduced Stress and Better Habits:** When individuals are joyful and content, they are less likely to seek fleeting happiness through impulsive purchases or unnecessary splurges, leading to more mindful spending and saving habits.

- **Increased Productivity and Creativity:** Happiness and a positive outlook can fuel motivation and productivity, contributing to better work engagement and financial results.

- **Generosity and Connection:** Joy can inspire generosity, which in turn can deepen happiness and foster a sense of fulfillment.

2. Service and wealth creation

- **Value Creation:** Wealth, in a spiritual sense, is often seen as a byproduct of providing value to others. By focusing on how one can serve and benefit others, individuals can create opportunities for themselves while positively impacting society.

- **Building Relationships and Trust:** A genuine attitude of service can build trust and stronger relationships with clients, colleagues, and community members, which in turn can lead to increased income and opportunities.

- **Positive Karma and Legacy:** Giving back to the community and engaging in charitable acts can enrich others' lives and potentially pave the way for one's own financial prosperity, notes beyondhappiness.love. Many wealthy individuals are driven by a desire to leave a positive legacy through philanthropy and wise wealth management.

3. Gratitude and wealth creation

- **Shifting from Scarcity to Abundance:** Gratitude cultivates an abundance mindset, shifting focus from what is lacking to appreciating what is present. This can lead to breaking free from fear-based thinking and approaching finances with greater confidence and positivity.

- **Enhanced Decision-Making:** Gratitude can reduce stress and anxiety related to finances, promoting clearer thinking and more mindful and intentional spending and saving decisions.

- **Increased Resilience and Patience:** By reinforcing a positive mindset, gratitude can build resilience, helping individuals to navigate financial challenges with greater confidence and patience, even during difficult market conditions.

- **Attracting Opportunities:** A grateful and positive mindset can radiate positivity, potentially attracting new opportunities and fostering beneficial relationships that can contribute to financial growth.

Sacred Prosperity Principles

These concepts of joy, service, and gratitude are woven into many "sacred prosperity principles" found in spiritual traditions and personal development teachings. Key themes include:

- **The Law of Attraction:** Suggests that positive thoughts and emotions (including joy and gratitude) can attract positive outcomes, while negative thoughts attract negative ones.
- **Mindfulness and Intentionality:** Emphasizes living in the present moment and making conscious choices that align with one's values and purpose. This can involve setting clear intentions for financial goals and aligning those intentions with spiritual and personal values.
- **Stewardship:** The idea of managing resources responsibly and using them to create positive impact, not just for oneself, but for the greater good.
- **Generosity and Giving Back:** Recognizing that giving back to the community and contributing to the well-being of others can create a positive cycle of prosperity and fulfillment.

By embracing these principles, individuals can aim to build a financial foundation rooted in values and purpose, attracting not just material wealth but also a sense of inner peace, fulfillment, and a legacy that benefits both themselves and the world around them.

Prosperity is not defined by a pricetag.

Chapter 13: Restoring Harmony with Earth and Cosmos: Forgotten Currents, Ley Lines, Love Grid Lines, Chakras

The Sacred Conclusion – Why This Chapter Ends the Book

I have chosen to close this book with a return to the Earth—not metaphorically, but quite literally. Over the years, through my devotion to yoga, sacred teachings of Jesus, and a deepening relationship with the natural world, I've come to understand that the human journey is inseparable from the Earth's own energetic blueprint. What we often describe as spiritual awakening is, in many ways, a reconnection to the frequencies of the planet itself—its rhythms, its wisdom, and its memory stored in the invisible lines that run beneath our feet.

These lines are not fantasy. They are as real as the circulatory system within your body. Known in ancient traditions as **ley lines**, **dragon lines**, or **songlines**, they are the acupuncture meridians of Mother

Earth. Just as we experience blocked energy in our physical and emotional systems, the Earth too can suffer from stagnation, interference, or trauma in her energetic grid. And just as we heal through touch, sound, frequency, and light, so too can she.

Ending this book here is not an ending at all—it's a homecoming. For if we are to remember the God Spark within, we must also remember our connection to the living library of this planet. The ancients knew this. From the Egyptian pyramids to the Hopi mesas, from Stonehenge to the temples of Peru, sacred sites were built with reverence to **celestial alignments**, **longitude and latitude crossings**, and **stellar gateways**. These locations weren't chosen at random—they were the *intersections of Heaven and Earth*, encoded with divine mathematics.

As you read this chapter, I invite you to feel the pulse of the planet as your own. Imagine yourself as a living antenna—anchored in Earth, reaching toward the stars, conducting frequency from both. This is not poetic fancy. This is **sacred science**, and it is time we remembered.

Healing the Earth's Energy Grid: Ley Lines and Sacred Geometry

Ley lines are the arteries of Earth's subtle body. Like the meridians that flow through our physical forms, these energetic highways carry life force—chi, mana, prana—across continents. They intersect at nodal points of power, often where temples, pyramids, stone circles, or cathedrals now stand. These alignments are not coincidental; they reflect a deep understanding of energy, geometry, and celestial placement.

Sites like the Great Pyramid of Giza, Stonehenge, Machu Picchu, Uluru, Teotihuacan, Glastonbury Tor, and Sedona were intentionally built at these intersections. Why? Because the ancients knew how to read the Earth's pulse. They observed the stars, studied the land, and felt the vibrational shifts beneath their feet. They mapped these sacred geometries in alignment with solar solstices, equinoxes, planetary transits, and even the procession of the equinoxes across ages.

Each sacred site was a tuning fork—aligning heaven and earth, human and divine. These places are still humming with encoded frequencies. But over time, the flow has been obstructed by war, pollution, electromagnetic interference, and spiritual neglect.

Today, Sacred Site Activation involves reawakening these points—by anchoring love, breath, prayer, sound, and intention back into the grid. Just as acupuncture rebalances the body's chi, this work re-patterns the Earth's field.

Love Grid Lines and Planetary Ascension Currents

While ley lines are ancient and structural, "love grid lines" are higher-frequency pathways that support the heart of the planet and its spiritual evolution. These currents—some call them Christ lines, Magdalene lines, or Unity grids—are associated with compassion, truth, healing, and the divine feminine principle.

Where ley lines form the bones of the Earth's energetic body, love lines are its breath and blood—fluid, radiant, and deeply attuned to human consciousness. Many mystics believe these lines have become more active as humanity enters a higher vibration, and as Earth herself ascends into a new octave of awareness.

Places like Chartres Cathedral, Mount Shasta, the Mary and Michael lines in Britain, and Taos Mountain in New Mexico carry the resonance of these love frequencies. When we connect to them with open hearts, we assist in the repair and amplification of these divine channels.

Human Chakras and Earth Chakras: A Mirror of Divine Design

Just as the human body contains chakras—spinning vortexes of energy that regulate our physical, emotional, and spiritual balance—the Earth also has chakras. These planetary chakras are large, high-frequency nodal points located across the globe, corresponding to major intersections of ley lines and sacred sites.

Each Earth chakra corresponds with a human energy center:

- **Root (1st):** Mount Shasta, California
- **Sacral (2nd):** Lake Titicaca, Peru/Bolivia
- **Solar Plexus (3rd):** Uluru-Kata Tjuta, Australia
- **Heart (4th):** Glastonbury & Shaftesbury, England
- **Throat (5th):** Great Pyramid of Giza, Egypt
- **Third Eye (6th):** Mobile chakra (linked to various sites, especially Europe/Middle East during shifts)
- **Crown (7th):** Mount Kailash, Tibet

When we travel to, meditate with, or send energy to these locations, we are helping activate and realign the Earth's chakra system. This also realigns our own, creating a harmonic resonance between body and planet, soul and cosmos.

Longitude, Latitude, and Star Alignment: Earth's Position in the Cosmic Symphony

Ancient navigators and sacred architects didn't just map the land—they mapped the sky. Longitude and latitude were more than geographic coordinates; they were harmonic placements in the cosmic orchestra.

Every sacred site is position batteriesned in celestial alignment with constellations, star portals, and galactic centers. The pyramids align with Orion's Belt. Angkor Wat mirrors Draco. The Nazca lines in Peru depict animals only visible from above—as if communicating with the stars. Even the Mayan calendar and the stone temples of Chichen Itzá were calibrated to solstices, equinoxes, and Venus cycles.

Why? Because the ancients knew we are not isolated. We are stardust in motion. Light from Sirius, pulses from the Pleiades, and frequencies from the Galactic Center affect our biology, consciousness, and collective memory.

When we realign with the Earth's geometry, we also realign with the heavens.

Ancient Tools, Living Intelligence – Divining Rods and the Invisible Currents of Earth

If the Earth has a system of energy lines, how do we find them? This is a question both ancient mystics and modern explorers have asked for centuries—and the answer has often come not from technology, but from the Earth herself.

One of the oldest and most astonishing tools of energy detection is the **divining rod**, also known as a dowsing rod. Traditionally crafted from Y-shaped branches or two bent metal rods, these tools are held lightly in the hands, and when the person passes over a subtle energetic current—whether water, mineral veins, or ley lines—the rods move, crossing or separating in response to the invisible flow.

To the untrained eye, this may seem like superstition. Yet dowsing has been used by farmers, builders, and spiritual seekers across nearly every continent. Entire cities—including parts of Europe and Indigenous villages—were constructed based on the confirmation of energetic harmony through this method. Even **the Vatican** and **the military** have employed dowsers to locate underground water and tunnels. Why? Because **the Earth is alive**, and these rods simply serve as an extension of our own bioelectric sensitivity.

When I first witnessed this phenomenon, I was amazed. No batteries. No screens. Just the subtle movement of a rod responding to an unseen truth. What struck me even more deeply was this: the human nervous system reacts the same way. Like the rod, we are built to *feel* alignment and misalignment, to sense what is unseen. Dowsing rods simply make it visible.

This chapter, then, is not about mysticism for mysticism's sake—but about **remembering our place inside an intelligent design**, where tools like divining rods reveal the subtle truths that our ancestors never forgot.

Migration, Memory, and the Cosmic Compass

Long before humans drew maps or spoke of longitude and latitude, the Earth's creatures were already navigating by a **higher intelligence**—one that science is only beginning to understand. Birds migrate thousands of miles, year after year, often returning to the exact same tree or shoreline. Whales travel along ancient underwater routes that mirror land-based ley lines, singing to each other across oceans. Sea turtles hatch on a beach and return decades later to the very same stretch of sand to lay their eggs.

How?

The answer lies not in GPS, but in **geo-magnetic resonance**, **celestial alignment**, and an inborn **connection to the Earth's energetic grid**. Animals tune in to something we have forgotten—something that was once as natural to us as breath.

These migrations are not random. They follow paths of least resistance and highest coherence, the same way rivers find valleys and wind finds open space. In many esoteric traditions, these paths are understood to be **ley lines** or **love lines**—currents of energy that carry both biological instinct and spiritual memory.

Indigenous peoples understood this well. Aboriginal Australians mapped their sacred landscape through **Songlines**—vibrational tracks passed down through stories and ceremony. Native tribes across the Americas followed the "spirit paths" of buffalo and eagle, believing these creatures knew how to follow the lines laid down by the stars.

Long before scientists place seismographs or issue tsunami alerts, animals often move first. Snakes, sensitive to vibrations through the ground, will abandon their burrows and slither to safer ground days before an earthquake. Birds suddenly fall silent or take flight in mass exodus. Elephants and dogs grow restless and head inland before a tidal wave hits. Even grazing herds have been observed shifting toward higher terrain hours before rising waters. These responses are not coincidence—they are the language of resonance and attunement to the Earth's subtle signals. Where human instruments measure data after it begins, animals perceive danger as a felt vibration in the field of life itself. Their behavior reminds us of a deeper intelligence that modern people often overlook: the body's capacity to sense and respond to the unseen currents of change.

Modern science may call it instinct. I call it **remembrance**.

When we restore harmony with these currents—whether through pilgrimage, prayer, planting, or presence—we begin to remember that we too are migratory beings. We are wired to follow light, to align with flow, to respond to the call of something greater than ourselves. Something unseen, but felt.

Just like the whales.

Just like the stars.

Practical Restoration: Harmonizing the Grid Within and Without

This sacred work is not reserved for monks, mystics, or those hidden in temples. It is for you. It is for all of us. Every human being, regardless of background or belief, carries within them the capacity to become a **living tuning fork**—an instrument of coherence, transmitting peace, healing, and divine order wherever they walk.

You do not have to cross oceans or climb mountains to shift the frequency of the world, though pilgrimages to sacred sites do amplify the signal. But the truth is: the grid runs through **you**. You are standing on holy ground, right now.

To restore harmony within and around you, begin where you are.

You can:

- **Walk barefoot on the Earth**, allowing your body to recalibrate through direct connection with the planet's subtle electrical charge. This is not just grounding—it is remembrance.

- **Offer a song, a prayer, or a whisper of gratitude** to the trees, the stones, the wind, or a nearby mountain. Spirit listens. Matter listens.

- **Work with sacred tools**—crystal grids, essential oils, tuning forks, geometric symbols, anointing oils, holy water, salt, fire—infused with intention.

- **Meditate while seeing the Earth wrapped in golden light**, her ley lines pulsing in soft, luminous rhythm beneath your feet.

- **Ask to be shown the energy beneath your home, town, or path**. Then bless it. Walk it with love. Trace the lines with your footsteps and presence.

Even the smallest act—lighting a candle with reverence, humming a healing note in the wind, or offering a smile that springs from soul-truth—creates harmonic resonance. **You are not separate from the map. You are a node on the grid.**

Jesus spoke of this directly, though few recognized the science hidden in his teachings. In Love Without End: Jesus Speaks, he said:

"The heart's power is rectified through the way it connects body and soul."

You are the bridge between worlds. The intersection point of spirit and substance. When you move in love, in truth, in coherence—your very presence **activates dormant codes in the environment** around you. Your light blesses the land.

He also said:

"Whenever you experience the soul, you are restored to your original state."

And again:

"The heart is not merely a metaphor. It is the frequency center of creation, and its attunement determines the direction of your whole being."

So to harmonize the grid of Earth, begin by harmonizing your own frequency. **A peaceful heart becomes a lighthouse. A grateful mind becomes a conductor of miracles. A loving presence becomes a map correction in the field.**

You do not need to fix the world.

You need to tune yourself.

And the grid will respond.

A Call to Sacred Stewardship

We are the remembering species. The carriers of ancient knowledge, encoded in our DNA and hearts.

To restore harmony with Earth and Cosmos is to remember that we are co-creators of the sacred dance. The breath of Gaia lives in our lungs. The light of Sirius reflects in our eyes. The magnetic pulse of the core flows through our spine.

Let this chapter be your invitation—to walk as a tuning fork of love, a keeper of coherence, and a steward of planetary awakening.

We do not restore harmony through striving, fear, or control.

We restore harmony through remembrance, alters of all types and devotional practices.

We are connecting to the song the Earth has been waiting to hear again.

Final Activation: Sacred Sites, Healing HeARTS Love Grid Prayer

Place your feet flat upon the Earth, or close your eyes and imagine you are standing on sacred ground—because you are. Breathe in deeply, and speak these words aloud or within:

I remember now.
I am a living part of Earth's great design.
I am not separate—I am sacred.
My body is a bridge, my breath is a blessing,
my heart is a transmitter of holy frequency.

I tune myself now to divine coherence.
May the light of Christ within me
illuminate every cell, every thought, every word.
I surrender distortion, fear, and illusion.
I align with love, truth, and sacred responsibility.

As I walk, may the land beneath me remember.
As I speak, may the waters respond.
As I sing, may the stars rejoice.
As I pray, may the grid realign.

I bless the ley lines and love lines.
I bless the healing of all sacred sites,
past, present, and future.

I stand as a node of light in the Great Map.
I remember who I am.

I open myself to God working in and through me, AS me.

In the name of the One Frequency of Love,
I activate the Spark of God within me.
And so it is.
And so I shine.

STARSHINE

by Marshall James Kavanaugh
Written live at the Taos Farmers Market, August 2025
@dreampoetforhire | Taos, NM

gritty roots
that uplift the most
radiant flower
transmuting through
the illumination
of a solar star
that shines

this precious peace
and glory to abundance
a love permeating through
the earth
all the way to divinity

there are ways
that we each
have a light
that we can beam

rising from the earth
a new dream
starts from the garden

it grows in the heart's echo
and opens wider

is horizon
to something that is sacred
as a constellation
and all its mythos

may we each
connect our rhythm
and find the brightness
no matter the hour.

Marshal Reads Starshine After He Wrote It

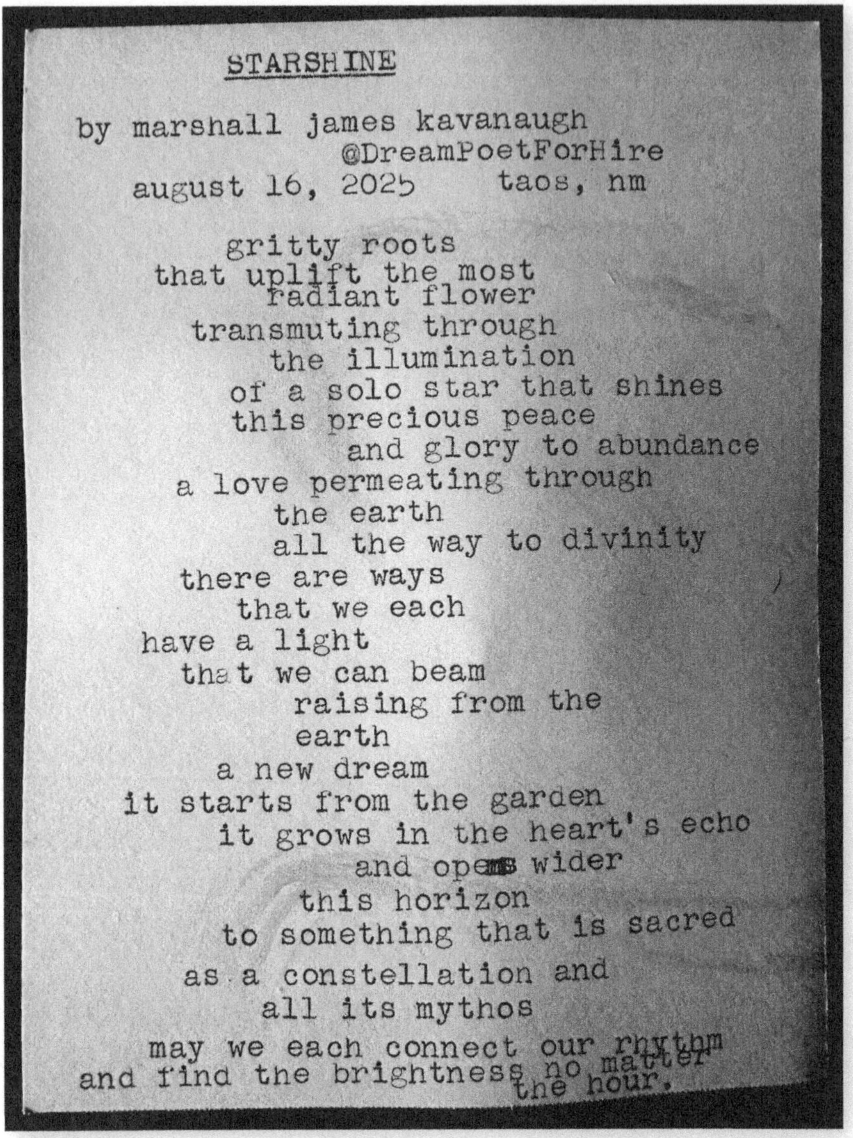

STARSHINE

by marshall james kavanaugh
@DreamPoetForHire
august 16, 2025 taos, nm

gritty roots
that uplift the most
radiant flower
transmuting through
the illumination
of a solo star that shines
this precious peace
and glory to abundance
a love permeating through
the earth
all the way to divinity
there are ways
that we each
have a light
that we can beam
raising from the
earth
a new dream
it starts from the garden
it grows in the heart's echo
and opens wider
this horizon
to something that is sacred
as a constellation and
all its mythos
may we each connect our rhythm
and find the brightness no matter
the hour.

About the Author

Dr. Trish Sanders, Ph.D.

Metaphysician · Author · Speaker · Minister · Entrepreneur

Dr. Trish Sanders—widely known as *Dr. Trish*—has spent her life bridging ancient wisdom and modern science to help people awaken to their fullest potential. She began her journey teaching yoga at age 17, self-taught from the groundbreaking television series by Lilias Folan, and soon after studied under an Iyengar Master in San Diego, where she earned her first Yoga and Pranayama certification. As part of her training, she made a sacred vow to teach for life—a promise she continues to fulfill to this day.

Her career spans decades of devotion to personal and planetary transformation, weaving together spiritual teaching, corporate leadership, and educational innovation. As founder of **StarShine Enterprises**, she has opened schools worldwide, trained teachers, and developed programs rooted in holistic well-being—body, mind, spirit, health, wealth, and happiness.

Dr. Trish holds advanced certifications in hypnosis, aromatherapy, sound therapy, Reiki, and Merkabah activation. Her Master's thesis, *Energy Medicine and Taming the Brain with God*, and her Ph.D. in Energy Medicine focus on using frequency and consciousness as foundational tools for building regenerative, soul-aligned businesses.

Her work has brought her into collaboration with global spiritual and scientific leaders—including Pope Francis—through fellowships with Arizona State University, Windsor Castle, and appearances at the United Nations, the Pontifical Academy of Sciences, and the International Space Conference. She also served for 23 years as the resident yoga teacher at the renowned Phoenician Five-Star Resort and has taught in luxury spas and retreats around the world.

Now based in Phoenix, Arizona, Dr. Trish continues to lead transformational retreats and projects—including the development of the *STARSHINE.life* app, the *FlowZone.world* movement, and the *Sacred Sites Healing HeARTS Revival Festivals*—to elevate human consciousness, awaken remembrance, and restore harmony on Earth.

Made in the USA
Monee, IL
05 November 2026

67375742R10104